THE PREPPER'S WATER SURVIVAL BIBLE:

THE ULTIMATE GUIDE ON HOW TO FIND, HARVEST, FILTER, PURIFY, AND STORE WATER OFF THE GRID

DALE MANN

FREE GIFT
TO OUR READER

Emergency water survival guide

PLEASE SCAN THE QR CODE BELOW
TO ACCESS THE BONUS MATERIALS.
THIS WILL DIRECT YOU TO GOOGLE
DRIVE. IF YOU DON'T HAVE GOOGLE
DRIVE, PLEASE DOWNLOAD IT FOR
FREE.

TABLE OF CONTENTS

INTRODUCTION

Imagine you're hiking in a scorching desert, miles from anywhere. You reach for your water bottle - empty! What started as an adventure turns into a fight for survival. Out here, water isn't just a drink, it's the difference between life and death.

This book isn't about some crazy desert scenario (though that sure highlights how important water is!). It's about the real deal: why water is so critical when things get tough.

My travels and experiences, from beautiful landscapes to challenging situations, have all taught me one thing - water is key. That's why I've spent years learning how to find, clean, and store water, no matter where you are. But even more importantly, I want to share this knowledge with you.

Whether you're new to exploring or a seasoned prepper, this book is your guide to getting reliable water, anywhere, anytime. "The Prepper's Water Survival Bible" is like a roadmap to managing this essential resource. It's written for adults who are serious about survival and teaches the skills you need to find water even in dry places.

The book covers everything from simple ways to collect water to advanced methods for cleaning and storing it safely. You'll learn how to catch rainwater, turn dirty water drinkable, and even create hidden water stashes. It's packed with practical tips, DIY projects, and true stories to make the information clear and useful.

What makes this book different? It focuses on sustainable and legal ways to get water, considers the environment, and includes the latest water purification tech. With emergencies becoming more common, knowing how to manage water is vital.

This book isn't just for reading, it's for taking action! See how prepared you are for water shortages and use this book as a guide to improve. The skills you learn here aren't just knowledge, they could save your life.

Through my own close calls and successes in the wild, I learned why being prepared is so important. It's about not just surviving, but thriving, no matter what comes your way. This book is my way of sharing those lessons. Together, let's be ready for anything and become masters of our own survival.

Welcome to "The Prepper's Water Survival Bible" - your path to water independence starts now!

CHAPTER 1
WATER AND THE EFFECTS OF NOT HAVING ANY

O ur bodies are amazing machines, but just like a car on a hot day, they need water to keep running smoothly. When you don't drink enough fluids, dehydration sets in, and it can make you feel crummy.

This chapter dives deep into what happens inside your body when you're dehydrated. It explains the sneaky signals your body sends out when it's craving water, like secret messages. By learning these signs and taking action before you feel terrible, you can avoid the problems that come with forgetting to drink enough fluids. It's all about understanding your body's way of saying "help, I'm thirsty!"

1.1 DEHYDRATION: RECOGNIZING THE WARNING SIGNS AND STAYING HYDRATED

Dehydration occurs when the body loses more fluids than it takes in, disrupting its normal functions. Our bodies, composed of about 60% water, rely on this essential fluid for everything from transporting nutrients to eliminating waste. When water levels drop, this delicate balance is thrown off. Ignoring dehydration can quickly turn it from a manageable issue into a serious health threat.

Early Warning Signs: Your Body's Signals

The body isn't shy about communicating its thirst. The most well-known sign is thirst itself, a primal urge prompting us to replenish lost fluids. But thirst isn't the only clue. A dry mouth, that uncomfortable feeling when talking or swallowing, is another indicator of low water levels.

Another way to gauge hydration is your urine color. When well-hydrated, urine is typically a light straw color. As dehydration sets in, the body concentrates waste products, turning the urine a deeper amber. This change is your body's way of conserving water. These early signs, though subtle, are crucial messages. Heed them to prevent dehydration from progressing.

Severe Dehydration: When the Signs Become Urgent

If you continue to ignore your body's pleas for hydration, dehydration intensifies. The symptoms become more pronounced and alarming. Confusion sets in as your brain, largely composed of water, struggles with the lack of hydration. Your heart also feels the strain, working harder to maintain blood flow as circulating fluids diminish. Fainting or syncope might even occur due to reduced oxygen reaching the brain. These severe symptoms are like blaring alarms, demanding immediate action.

Prevention is Key: Staying Ahead of the Curve

The good news is, dehydration is entirely preventable. The key lies in proactive hydration, not just reacting when thirst strikes. Regularly sipping water throughout the day, even when you don't feel parched, is essential. Think of it as deliberately maintaining your body's fluid levels, rather than waiting for a warning sign. This is especially important in situations that promote rapid fluid loss, like hot weather, high humidity, or during exercise.

Don't underestimate the power of your diet either. Water-rich fruits and vegetables can significantly contribute to your daily fluid intake. In our fast-paced world, prioritizing hydration might seem like a minor detail, but the consequences of neglecting this basic need can be serious. Dehydration is a stark reminder that water isn't just a drink; it's the lifeblood of every cell in our body. By honoring this fundamental need, we empower ourselves to face any challenge with resilience.

1.2 KNOW YOUR DAILY WATER NEEDS

Our bodies are like intricate machines, constantly whirring and working to keep us alive. Water, the essential lubricant of this machine, keeps everything running smoothly. But unlike a car that holds a fixed amount of fuel, our water needs are constantly changing. This chapter delves into the factors that influence how much water we need, empowering you to personalize your hydration strategy for optimal health.

Age: A Thirst for Growth

Children, those bundles of boundless energy, have a higher water requirement per kilogram of body weight compared to adults. This makes perfect sense. Their bodies are in a constant state of growth and development, demanding a steady flow of water to fuel these incredible changes. However, as we age, the water content in our bodies naturally decreases. This shift, combined with a potential dulling of the thirst sensation, makes it crucial for older adults to prioritize water intake, even when thirst isn't screaming.

Weight: A Matter of Balance

It's no surprise that someone who weighs more typically needs more water to function properly. However, it's not just about total body weight. The composition of that weight also plays a role. Muscle tissue, for example, holds more water than fat tissue. So, individuals with a higher muscle mass may find themselves needing slightly more water to stay hydrated. It's a beautiful testament to the intricate balance our bodies maintain.

Climate: A Silent Influence

The environment we live in plays a significant role in how much water we need. Imagine yourself basking under the scorching desert sun compared to being nestled comfortably in an air-conditioned office. In the desert, sweat becomes our body's primary cooling mechanism, leading to increased water loss through respiration as well.

Similarly, those living in tropical climates, where humidity hangs heavy in the air, will find their water needs amplified. The environment, in its quiet way, constantly influences how much water our bodies require.

Activity Level: The Rhythm of Hydration

Our daily activities, from the leisurely stroll through the park to a grueling workout at the gym, dictate the pace at which we use and lose water. Physical exertion, regardless of intensity, accelerates water loss primarily through sweat.

This loss is not just water, but also essential electrolytes, those tiny warriors that keep our muscles functioning and our nerves firing. So, when we sweat heavily, we need to replenish not only fluids but also these vital minerals.

General Guidelines: A Starting Point

Given the vast array of personal and environmental factors that influence water needs, general guidelines can serve as a helpful starting point. The U.S. National Academies of Sciences, Engineering, and Medicine suggests a daily intake of about 3.7 liters (approximately 125 ounces) of fluids for men and 2.7 liters (approximately 91 ounces) for women.

It's important to remember that these recommendations encompass all fluids consumed, including water from food and other beverages. Children, with their unique developmental needs, have guidelines that adjust with age, reflecting the ever-changing needs of their growing bodies.

Survival Situations: Adapting to Challenges

When faced with a survival situation, our water needs shift dramatically. Stress, both physical and emotional, coupled with the demands of navigating unexpected circumstances, can significantly increase water requirements. Think about it. If you're searching for shelter, gathering resources, or simply moving to safety, you're likely expending a lot of energy, leading to increased water loss through sweat.

Additionally, exposure to extreme weather conditions, be it scorching heat or freezing cold, further compounds this effect. In these moments, the body's call for water becomes more urgent than ever.

Planning for the Unexpected: Beyond Just Drinking

Planning for the unexpected isn't just a good idea, it's essential. Anticipating higher water needs in survival situations requires knowledge of how to find and purify water, as well as how to ration it effectively.

Rationing, a word that often conjures images of scarcity and struggle, is not about deprivation. It's about strategic allocation - ensuring that every sip serves your body's needs as effectively as possible. This involves understanding when and how much water you need to consume, but also incorporating water-saving strategies into other activities, like food preparation and personal hygiene.

Knowing about our individual water needs and how they are influenced by our bodies and environments, we lay the foundation for effective survival strategies.

Water may be abundant on Earth, but our bodies can't store it indefinitely.

Regular replenishment is vital. Moreover, both under-hydration and overhydration can have consequences. So, knowledge is important - It empowers us to navigate the challenges of both daily life and survival situations with confidence.

1.3 ASSESSING YOUR WATER NEEDS: INDIVIDUAL AND FAMILY PLANNING

How Much Water Do You Really Need? A Family Fun Guide (Skip the Math!)

Figuring out how much water you need can feel confusing, like trying to understand a kid's homework. But don't worry, water warriors! This guide will break it down for you, with less science talk and more "aha!" moments.

The Starting Point: Daily Needs

Imagine your water needs are like climbing a mountain. Everyone starts at the bottom, with a general rule of about a gallon of water a day. This is just a starting point, not a magic number. Think of it as the foundation for your personal "hydration house" (because houses are way cooler than pyramids!).

Level Up: Water Needs for You and Your Crew

Now, let's add some things to your house. Age, weight, how active you are, and even the weather all affect how much water you need. A little kid on a walk needs less water than a grown-up working out hard, just like you wouldn't wear a winter coat to the beach (hopefully!).

Don't Forget the Furry (or Feathery) Friends!

Our pets are family too, and they need water to stay happy and healthy! From Fido the water-loving dog to Mittens the surprisingly energetic cat, each furry (or feathery, or scaly) friend needs a different amount of water. Even Sparky the hamster needs his share! Thinking about your pet's water needs is like packing for the whole family on vacation – everyone needs to stay hydrated for a fun trip.

Planning for the Unexpected: Beyond Daily Needs

What if it doesn't rain for a long time? Planning for emergencies is like packing for an unexpected adventure. You wouldn't just throw some clothes in a bag and hope for the best. You need to figure out how much water to store, where to keep it, and how to make sure it's safe to drink. Even in an emergency, you don't want to be stuck drinking pond water (unless you're super lost, then good luck!).

Be Flexible: Life Can Be Surprising!

Life loves to throw curveballs. Sometimes there's no rain, other times pipes burst, and sometimes your teenager forgets their water bottle (again!). The key is to be able to adjust. Being able to change how much water you drink and how you store it based on what's happening is like having a Swiss Army knife on your hydration hike. It helps you handle whatever unexpected challenges come your way.

Knowledge is Power: Hydration Hacks!

Just like having a map helps you on a hike, knowledge is your best friend when it comes to water. Learning how to purify water, store it safely, and even how to use it wisely (think shorter showers and fixing leaky faucets) are all great skills to have. These are your hydration hacks, the cool tricks that make your journey smoother and more fun.

Sharing is Caring: Building a Hydration Team

Remember, nobody is an island (especially when it comes to water!). Sharing what you know about water with others is like creating a hydration team. By working together, we can all be better prepared and make sure everyone has access to this life-giving resource.

Figuring out how much water you need doesn't have to be boring. A bit of planning, flexibility, and some teamwork, you can become a hydration pro, keeping yourself and your loved ones healthy and happy, no matter what life throws your way.

1.4 FINDING WATER: A PREPPER'S OVERVIEW

Water, the stuff that keeps us alive, becomes the most important thing to find if you're ever in a situation where you're on your own. This guide will help you understand the different places you can find water, the good and bad things about each place, and how to get the water safely.

Nature's Bounty: A Gift with Hidden Dangers

The most obvious places to look for water are in nature: rivers, lakes, and springs. These can be great sources of water, but there are a few things to keep in mind. The amount of water flowing in a river or stream can change depending on the season.

In the summer, there might be a lot of water, but in the winter, the river might be just a trickle. Also, pollution from things like storms or factories can make the water unsafe to drink. The key is to learn as much as you can about the area you're in and how the water flows there.

Urban Oasis: Rainwater and Unexpected Reserves in Cities

Even in concrete jungles like cities, there can be hidden reserves of water. Rainwater, which most people just let fall to the ground, can actually be collected from rooftops and stored in containers. Public fountains and fancy water features in parks, while not the best for everyday use, can become emergency reserves if you treat the water properly.

The problem with city water sources is that dirty air and chemicals used in the city can get into the water, so you'll need to clean it before you drink it.

Tapping the Depths: Finding Water Underground

The way you find water can depend on the kind of land you're in. In dry areas, for example, people rely on underground reserves of water called aquifers.

To get water from these aquifers, you'll need special tools that are different from the tools you would use to get water from a lake or river. Climate also matters. If you're in a place that doesn't get much rain, you'll need to use different strategies to find water than someone who lives in a place that gets a lot of rain.

Sustainability: Using Water Wisely

As we take water from these different sources, it's important to remember that water is not something that lasts forever. We need to find a way to use water without hurting the environment.

This means wasting less water, not polluting the water we use, and finding ways to bring back the water we take. By using water wisely, we show that we understand how important water is for all living things, not just ourselves, and that we care about future generations having clean water to drink.

The Prepper's Challenge: More Than Just Finding Water

Being prepared for anything isn't just about finding water to drink right now. It's also about finding ways to get water in the future without hurting the environment.

You'll need to know how to find water, but you'll also need to understand how to use water wisely. This way of thinking about water shows that you're not just looking out for yourself, but for your community and the environment too.

A Journey of Discovery: Understanding Our Dependence on Water

Finding water isn't just about having something to drink. It's a chance to learn more about the world around us. When we look for water, we start to see how important water is for everything that lives. We see the delicate balance of nature and how much we depend on it. This journey reminds us that we need to take care of the environment, which is a big part of being prepared for anything.

Responsibility Beyond Our Needs: Thinking About the Future

The choices we make about finding water today can affect the world tomorrow. If we waste water or pollute it, it will be harder for people and animals to find clean water in the future. By being aware of this, and by using water wisely, we can find ways to get the water we need without hurting the environment. This means thinking about not just ourselves, but about everyone who will come after us.

CHAPTER 2
UNVEILING HIDDEN HYDRO ASSETS: URBAN WATER MASTERY

Cities, with their towering buildings and busy streets, might seem like dry places with no water to be found. But that's not quite true! Water is all around us, hidden in plain sight, just waiting to be collected. This chapter will show you how to find and use this "hidden water" in an urban environment.

2.1 URBAN WATER SOURCES: A HIDDEN BOUNTY

Cities have many places where water collects that we might not think about. One big source is rooftops. When it rains, the water runs off the roof and goes down the drain. But with a little work, you can set up a system to catch this rainwater and store it for later use.

Air conditioners are another surprising source of water. As they cool the air inside your home, they also take out moisture. This moisture condenses and drips out as water, which you can collect.

Even the hot water from your heater can be a source in an emergency, although it's important to check if it has any chemicals added that could make you sick.

Safety First: Keeping Your Water Clean

The water you collect in the city can be dirty. Bird droppings, pollution from factories, and other things can get in the water and make you sick. To make sure the water is safe to drink, you'll need to clean it.

- Air conditioner water: This water needs to be filtered and then boiled to kill any germs.
- Rooftop water: You might need to filter this water as well, especially if there are factories or busy streets nearby. You might also need to take out any chemicals that washed off your roof.

Knowing the Rules: Legal Considerations

Every city has its own rules about collecting rainwater and other water sources. Some cities allow you to collect a lot of water, while others only allow a little. It's important to check the laws in your area before you start collecting water.

Here are some examples of how rainwater collection laws can vary:

- In Colorado, you can only collect rainwater in two barrels that hold 55 gallons each.
- In Georgia, you need a special permit to collect rainwater for anything other than watering your plants.

The best way to find out the laws in your city is to call your local government office or search online for information.

Clever Collection: Making the Most of Your Space

Cities don't have a lot of extra space, so you need to be creative when collecting water. Here are a few ideas:

- Rainwater harvesting systems: These systems can be simple or complex. A simple system might just use buckets or barrels to catch rainwater runoff from your balcony or window ledge. A more complex system might involve changing your gutters so they collect the water and send it to a big storage tank.
- Modular water storage: These are special containers that are designed to fit into small spaces. They can hold a lot of water without taking up much room in your apartment or house.

Checklist for Safe Urban Water Collection

Here's a checklist to help you get started collecting water safely in the city:

1. Find your water sources: Make a list of all the places you might be able to collect water in your area, like rooftops, air conditioners, and heating systems.
2. Check for contamination: For each water source, think about what might make the water dirty and how you'll clean it.
3. Know the law: Research the rules about collecting water in your city.

4. Design your system: Figure out how you'll collect the water, considering how much space you have and what resources you have available.

5. Clean your water: Decide how you'll make the water safe to drink, like filtering and boiling.

6. Store your water: Find containers to store your water that won't take up too much space and will keep the water clean.

With these steps, you can turn the concrete jungle into a source of your own water! Cities are full of hidden water, waiting to be discovered by those who know how to look. With a little knowledge and some planning, you can be prepared and have all the water you need, even in the middle of the city.

2.2 RURAL WATER SOURCING: NATURE'S BOUNTY AND SUSTAINABLE PRACTICES

The vast countryside, with its rolling hills, meadows, and forests, offers a unique set-up for sourcing water. Unlike the urban jungle, resources are readily available in the form of streams, lakes, and the often-overlooked gift of rainwater. However, tapping into these natural sources requires not just knowledge and skill, but also a deep respect for the delicate balance of the rural ecosystem.

Accessing Natural Water Bodies: Respecting the Source

Streams, lakes, and ponds are tempting sources of water, but how we approach them matters. The first step is to minimize any disturbance to the surrounding area.

This means using existing pathways instead of creating new ones, leaving the vegetation untouched, and staying a safe distance from the water's edge to avoid causing erosion. When collecting water, use clean containers and draw from flowing sections of streams or rivers whenever possible. Stagnant water is more likely to be contaminated.

Rainwater Collection: A Sustainable Solution

Rural settings are ideal for collecting rainwater, a practice that aligns perfectly with sustainable living and self-sufficiency. Rainwater harvesting systems can be as simple as placing barrels under downspouts or as elaborate as systems with built-in filtration and storage tanks.

The key to maximizing efficiency is strategically placing collection containers to capture runoff from roofs, gutters, and other surfaces. Regularly cleaning and maintaining these systems ensures they function properly and stay clean. To further improve water quality, consider using first-flush diverters.

These clever devices divert the initial flow of rainwater, which might contain contaminants like dust or bird droppings, away from your storage tanks.

Assessing Water Quality: Don't Be Fooled by Appearances

The tranquility of a rural stream or the crystal-clear surface of a lake might be deceiving. To ensure your safety, it's crucial to assess the water quality before drinking it. Start with a visual inspection.

Look for clarity and the absence of surface scum or debris, both good signs. However, appearances can be misleading. For a more accurate picture, consider using basic field test kits. These kits, readily available at most outdoor stores, can detect the presence of harmful bacteria, nitrates, and other pollutants. While offering a preliminary assessment of water safety, they are not a substitute for a professional lab analysis.

Sustainable Practices: Preserving the Balance

Long-term access to clean water in rural areas relies on practices that support the health of natural water bodies and the surrounding ecosystem. The core principle is to take only what you need.

Excessive water extraction can deplete natural resources faster than they can be replenished. Another crucial aspect is to avoid contaminating water sources. This means keeping chemicals, waste, and other pollutants away from streams, lakes, and ponds. Be mindful of runoff from your property and use biodegradable, non-toxic products near water bodies.

When it comes to rainwater harvesting, sustainability involves minimizing water loss through evaporation and leaks. Water-saving technologies like low-flow faucets and drip irrigation systems can be helpful here. Additionally, planting drought-resistant vegetation around your home can further reduce water demands.

By adopting these practices, you become a responsible steward of the rural environment, ensuring that the natural bounty of water continues to be a source of life for generations to come.

Remember, humans are just one thread. Our actions have the power to support the delicate balance of the ecosystem or disrupt it. By choosing sustainable practices, we choose a future where clean water flows freely, nourishing both ourselves and the natural world.

2.3 DON'T GET SICK! STAYING SAFE WITH WILD WATER

Nature beckons with its cool, inviting streams, shimmering lakes, and the promise of refreshing rainwater. But hold on a sec, adventurer! While these water sources might look beautiful and crystal clear, they can harbor hidden dangers that can turn your exciting trip into a tummy-trouble nightmare.

Tiny bugs, invisible chemicals, and even plain dirt can lurk in wild water, waiting to give you a case of the grumbles or something even worse. This chapter will equip you with the knowledge to spot these sneaky threats and ensure the wild water you drink is truly healthy and safe to swallow.

Not Always Crystal Clear: Unveiling the Troublemakers in Wild Water

We've all heard the saying "moving water is clean water," but that old adage can be misleading. Sure, a fast-flowing stream might seem like a better choice than a still pond, but it can still be teeming with microscopic nasties that can make you sick.

And just because the water looks sparkling clean doesn't mean it's safe to drink either. There's a whole world of tiny troublemakers you can't see with your naked eye, like bacteria, viruses, and parasites, just waiting to hitch a ride into your digestive system and wreak havoc.

Here's a deeper dive into the troublemakers you need to watch out for in wild water:

Microscopic Menaces: Imagine single-celled organisms so small you'd need a super-powered microscope to see them! These bacteria, viruses, and parasites can cause a variety of unpleasant illnesses, ranging from mild diarrhea and cramps to more serious problems like dysentery or giardia.

These tiny terrors can contaminate water through human or animal waste, decaying organic matter, or even contaminated soil runoff.

Hidden Chemical Hazards: These chemical contaminants can come from a variety of sources, some more obvious than others. Agricultural runoff from farms using pesticides and fertilizers can introduce harmful chemicals into waterways.

Industrial facilities might discharge pollutants like heavy metals or industrial byproducts. Even naturally occurring minerals like arsenic or fluoride, found in certain rocks and soil, can leach into water sources at unsafe levels. These chemicals can have both immediate and long-term health effects, depending on the type and amount ingested.

Dirt and Sediment Problems: While not as scary as the other troublemakers, dirt, sand, and other suspended sediment can still be a nuisance. They can clog up filters you might be using to clean the water, making the purification process less effective.

Additionally, high levels of sediment can make the water cloudy and unpleasant to drink.

The weather also plays a role in how much trouble these bad guys cause. Heavy rainfall events can wash more pollutants and contaminants from the landscape into streams and lakes. Conversely, drought conditions can concentrate existing contaminants in smaller volumes of water, making them even more potent.

Don't be fooled by appearances! Just because water looks clear and inviting doesn't mean it's safe to drink.

Cleaning Up Your Wild Water: Transforming it into a Safe and Healthy Beverage

Since wild water can be a breeding ground for these hidden dangers, it's absolutely crucial to clean it up before you take a sip. Here are a few ways to turn that potentially risky wild water into a refreshing and healthy beverage:

Boiling: This tried-and-true method is a simple and reliable way to kill most bacteria and viruses. All you need to do is bring the water to a rolling boil for at least one minute (or even longer if you're adventuring at high altitudes, where the boiling point is lower).

Imagine the water bubbling like a witch's cauldron - that's a good sign it's hot enough to kill the bad guys! However, boiling won't remove all contaminants, such as chemicals or parasites.

Water Purification Tablets or Drops: These are handy little packets you can carry with you on your outdoor adventures. They work by releasing chemicals that kill germs and viruses in the water, but they might not be effective against all types of microorganisms or remove any hidden chemicals.

Make sure to read the instructions carefully before using them, as some tablets require a specific waiting period after treatment before the water is safe to drink.

Filtration Systems: These are the ultimate water warriors in your wilderness backpack! They come in various shapes and sizes, and they use filters and sometimes even ultraviolet light to remove a wider range of contaminants from the water.

Mechanical filters trap sediment, while activated carbon filters absorb chemicals and improve the taste and odor of the water.

Some advanced filtration systems even incorporate UV light technology, which can kill bacteria, viruses, and parasites that might have passed through the other filters.

Don't skip the cleaning step! It's like putting on sunscreen before a day at the beach - it protects you from getting sick.

Learning from the Past and Looking to Science: Keeping You Healthy with Wild Water

People have been suffering from waterborne illnesses for centuries, all thanks to a refreshing gulp of untreated water. Imagine explorers from the past, weary from their travels, taking a big swig from a seemingly clear stream, only to be struck down by days of cramps and discomfort.

Thankfully, the days of such risky refreshment are behind us, thanks to advancements in science and our growing understanding of waterborne threats.

A Legacy of Waterborne diseases: Historical records are filled with tales of devastating outbreaks of diseases like cholera, typhoid fever, and dysentery, all traced back to contaminated water sources.

These outbreaks highlight the invisible dangers that can lurk in seemingly pristine water and the importance of proper sanitation and water treatment.

Science to the Rescue: Modern science has shed light on the complex world of waterborne pathogens and pollutants. Microscopic analyses have revealed the various types of bacteria, viruses, and parasites that can contaminate water.

Researchers have also studied the behavior and persistence of these contaminants in different environments, allowing them to develop more targeted treatment methods.

The Power of Purification Technology: Building on this scientific knowledge, water treatment technologies have become increasingly sophisticated. Simple methods like boiling and filtration have been refined and improved, while new technologies like ultraviolet light treatment and reverse osmosis offer even more effective ways to remove a broader range of contaminants.

Knowledge is Power: By learning from the past and embracing scientific advancements, we can make informed decisions about water treatment in the wild. This empowers individuals and communities to take charge of their health and well-being, ensuring access to clean drinking water on outdoor adventures and in areas with limited access to safe municipal water supplies.

The power of knowledge and science can keep you healthy and hydrated on your wilderness adventures!

Enjoying the Journey: Water Without Problems

So, the next time you're out exploring nature, remember: the sparkling stream or refreshing rainwater might be tempting, but take a moment to be a water warrior. By understanding the hidden dangers and using a little know-how, you can transform wild water into a safe and healthy companion on your outdoor adventures!

After all, the best part of any adventure is enjoying the journey, and staying healthy ensures you can create lasting memories without any unwanted stomach souvenirs.

2.4 INNOVATIVE URBAN RAINWATER HARVESTING TECHNIQUES

Cities are covered in buildings and pavement, leaving little space for rainwater to soak into the ground. This is where innovative rainwater harvesting techniques come in - they're like smart ways to catch rainwater in these concrete jungles. These methods use technology and clever design to turn cities into places that save water.

Modern Rainwater Harvesting Systems

Forget just sticking a bucket under a gutter! Modern systems are designed to fit right into our urban lifestyle. Imagine using your walls, balconies, or even windowsills to collect rainwater! These systems are often like building blocks that you can put together however you need, depending on your space. This means they can work in all sorts of homes, from tiny apartments to big complexes.

Buildings Designed for Rainwater Harvesting

Some buildings are planned from the start to include rainwater harvesting. These buildings might have special roofs called "green roofs" that are covered in plants and soil. These roofs soak up rainwater just like a sponge, and they also help keep buildings cool in the summer and warm in the winter. They're basically little green oases in the city!

Maximizing Rainwater Collection in Cities

To catch the most rainwater possible, city planners and architects are getting creative. Special surfaces can be put on buildings that help water run off into collection systems instead of evaporating away.

Additionally, there are smart water management systems that use sensors and weather forecasts to figure out the best times to collect water. These systems can even send extra water to places that need it, like community gardens or parks, making sure everyone benefits.

Keeping the Water Clean

Rainwater might seem clean, but it can pick up dirt and pollution from the air and from rooftops. That's why it's important to treat the water before using it. There are different ways to do this, depending on how you plan to use the water. First, filters take out big things like leaves and twigs.

Then, depending on how dirty the water is, it might go through special treatments like UV light or biological filters to get rid of even smaller icky stuff. With the right treatment, rainwater can be safe for many things, from watering plants to washing clothes (and even drinking, if it's treated very carefully!).

Real-World Rainwater Harvesting Success Stories

There are already lots of cool rainwater harvesting projects happening in cities around the world! In some places, whole communities have gotten together to turn their rooftops into giant collection systems.

This provides water for their gardens and helps to reduce the amount of rainwater that overwhelms sewage systems during heavy storms. In other places, buildings are designed with water management systems that capture and treat rainwater for things like toilets and cleaning, so they don't have to rely as much on city water supplies.

These examples show that rainwater harvesting can work in different situations and can be a big help for cities. As more and more people move to cities and the pressure on our water resources increases, using these techniques becomes even more important.

Rainwater Harvesting: A Beacon of Hope for Cities

In a time when the environment is changing quickly, cities are showing that they can be places of conservation. Rainwater harvesting is a perfect example of this.

It shows us that we can have a different relationship with water - not just taking it for granted, but seeing it as a precious resource that needs to be protected.

Instead of thinking of cities as places where water gets wasted, we can start to see them as places where we can collect and reuse water in smart ways. This shift in thinking paves the way for a future where everyone has enough clean water.

The Last Drop

Finding ways to manage water sustainably is both a challenge and an opportunity. It's a chance to reimagine our cities as places that are innovative, conserve resources, and bring communities together. Every drop of rain holds the potential for a new beginning.

CHAPTER 3

HARVESTING THE SKIES: RAINWATER COLLECTION REIMAGINED

In a world where technology often overshadows simple solutions, rainwater collection stands as a powerful reminder of human ingenuity. This age-old practice, still highly relevant today, blends necessity with creativity.

For those seeking self-reliance (preppers) and those passionate about environmental responsibility (sustainability advocates), capturing rainwater transforms a fleeting weather event into a dependable water source. It's a practice that combines traditional wisdom with modern innovation.

3.1 BUILDING A RAINWATER COLLECTION SYSTEM: A STEP-BY-STEP GUIDE

Rain, a gift from the heavens, offers an elegant and practical solution to water scarcity. However, capturing this resource requires thoughtful planning and execution. Building a rainwater collection system isn't just about saving water; it's a commitment to sustainability, a way to show respect for nature while meeting your daily water needs.

Design Considerations

When designing your rainwater collection system, several key factors need to be considered. The type of roof you have, whether metal, tile, or something else, will influence the quality and quantity of water collected. Metal roofs, known for their cleanliness and efficiency in channeling water, are ideal surfaces. On the other hand, roofs treated with chemicals or made of toxic materials should be avoided, or the collected water may not be suitable for all uses.

The size of your roof, also known as the collection area, directly affects how much water you can harvest. A simple calculation, taking into account your region's annual rainfall and the efficiency of your roofing material in collecting water, will give you an estimate of your potential water yield.

Finally, you need to consider storage capacity. This means balancing how much water you expect to use with the available space on your property. You want to have enough water to get through dry periods without having overflowing tanks that take up too much space.

Materials and Tools Needed

Assembling a basic rainwater collection system is surprisingly easy on the wallet. The core components you'll need are readily available and affordable. These include PVC pipes or gutters, filters for your downspouts, first-flush diverters, storage tanks (preferably made from food-grade materials), and mosquito screens to prevent insects from breeding in the stored water. As for tools, a drill, saw, wrenches, and sealant are commonly found in most households and should be sufficient for the installation.

Construction Process

Building your rainwater collection system happens in stages, each crucial for the system's efficiency and safety. First, gutters are attached to the edge of your roof, channeling rainwater towards a designated downspout. Here, a filter will remove debris like leaves and twigs.

A first-flush diverter is a valuable addition, as it ensures the initial, dirtier rainwater is diverted away from your storage tanks, keeping the overall quality of the collected water high. Finally, the clean water flows into storage tanks equipped with mosquito screens to prevent them from becoming breeding grounds for insects.

Positioning these tanks on a raised platform with a stable base allows gravity to do the work of moving the water, reducing the need for electric pumps.

For those living in urban areas or with limited space, rain barrels connected to existing downspouts offer a simpler yet effective alternative. This approach not only conserves water but also helps manage stormwater runoff, a major concern in densely populated areas.

Maintenance Tips

Maintaining your rainwater collection system is key to ensuring it functions properly and lasts for years to come. Regularly cleaning your gutters, downspout filters, and first-flush diverters prevents blockages and keeps the collected water clean. After heavy rain or storms, it's wise to inspect your storage tanks for leaks or cracks to prevent water loss.

Algae growth can be a challenge in stored water. To minimize this, consider placing your tanks in shaded areas and using opaque materials.

Rainwater Collection System Maintenance Checklist

- Gutter and Downspout Cleaning: Schedule cleanings twice a year to remove leaves and debris.
- First-Flush Diverter Inspection: Check and clean your diverter every three months to ensure it's working properly.
- Storage Tank Inspection: Perform monthly checks for leaks, cracks, and overall cleanliness.
- Mosquito Screen Check: Regularly inspect screens for holes or tears to prevent insect entry.
- Algae Prevention: Monitor water clarity and consider adding natural algae inhibitors if necessary.

This checklist serves as a guide for maintaining your rainwater collection system, emphasizing proactive measures to ensure clean, safe water.

With consistent care and attention, your system can become a reliable water source for years to come, reflecting a sustainable approach to hydration that respects the environment and harnesses the untapped potential of rainwater.

3.2: CATCHING THE WHISPER OF WATER: DEW AND FOG COLLECTION

Imagine a land where rain is a rare visitor, the sun beats down relentlessly, and the earth seems parched. Yet, in these seemingly desolate places, tiny droplets of water cling to leaves in the morning and hang in the air like wisps of smoke. These are dew and fog, and they can be lifesavers.

This section explores how we can capture this precious moisture, transforming the fleeting into the essential.

The Science Behind the Magic

The ability to collect dew and fog hinges on a simple scientific principle: condensation. When warm air cools down, it can no longer hold as much moisture. This excess moisture condenses, turning from an invisible gas into tiny water droplets.

At night, as the ground cools under a clear sky, the air near the surface reaches a special temperature called the dew point. This is the magic number where condensation happens! The cooler the surface, the faster this occurs, which is why dew forms on blades of grass and leaves long before it does on the ground itself.

Fog is like a cloud that hugs the ground. It forms when warm, moist air mixes with cool air, causing the water vapor to condense into tiny droplets suspended in the air. Just like with dew, these droplets can be captured if we know how.

The key to success lies in understanding what makes condensation happen best. Three things are important:

- Humidity: The amount of water vapor in the air. The higher the humidity, the more water there is to condense.
- Temperature: The cooler the air gets, the faster condensation happens.
- Nucleation Sites: These are tiny bumps or imperfections on a surface that water vapor can cling to and form droplets.

In places with little rain but high humidity, dew and fog collection techniques shine. They offer a way to get water without needing to dig wells or rely on unpredictable rainfall. All you need is the air itself and the cool touch of night, or the gentle caress of morning mist.

DIY Dew Catchers: Simple Tools, Big Results

If you're looking for a way to collect a small amount of water for your garden or even for drinking (after proper treatment of course!), building a dew collector from everyday materials is a great option. Here's the surprising part: the simpler the design, the better it often works.

Imagine a sheet of plastic or metal, tilted at an angle to catch the night's condensation. As the water vapor condenses on the cool surface, it trickles down and collects in a container placed beneath. The type of material you use matters. Surfaces that lose heat quickly, like metal, will cool down faster and collect more dew.

To maximize your catch, place your dew collector in an open area with good airflow, away from trees or buildings that might block the night sky. The cooler the collector gets, the more dew it will collect. Here's a neat trick: add a layer of absorbent cloth or a sponge at the base of the collector. This will draw even more moisture from the air and help it channel into your container.

Fog Nets: Capturing the Fog's Bounty

For communities or individuals with larger water needs, fog nets offer a more advanced approach. Imagine a giant net made of fine mesh, suspended between tall poles.

As fog drifts through the air, the tiny water droplets get caught in the mesh and condense into even bigger drops. These drops then roll down the mesh and collect in a storage tank at the bottom.

The placement and direction of the fog net are crucial. Ideally, the net should be positioned where the fog is thickest, and aligned perpendicular to the prevailing winds. This ensures the maximum amount of fog gets captured.

Building a fog net requires some planning and the right materials. Stainless steel or strong plastic mesh will last for years, while a sturdy frame ensures the net can withstand strong winds. The mesh itself needs to be a special size - fine enough to catch the tiny fog droplets but not so fine that it blocks the wind and makes the net act like a sail.

Just like your dew collector, regular maintenance is key. Cleaning the mesh to prevent clogging and checking the frame for damage will ensure your fog net keeps collecting water efficiently.

From Parched Lands to Oases: Real-World Examples

In some of the driest places on Earth, dew and fog collection projects are making a big difference. Take the Atacama Desert in Chile, for example. This vast, arid region receives almost no rain. But thanks to fog nets, life is flourishing.

Huge fog nets, stretching over hundreds of square meters, capture thousands of liters of water every day. This precious water is used for drinking, irrigation, and even helps to restore native plants.

Another inspiring example comes from Rajasthan, India, where villages struggle with dwindling groundwater supplies. Here, innovative dew harvesting structures have been introduced.

These simple systems capture the moisture from cool night air providing a vital source of drinking water and reducing the village's dependence on unreliable wells. The success of these projects demonstrates the power of dew and fog collection, transforming once-desolate landscapes into places where life can thrive.

Beyond Survival: The Future of Dew and Fog Collection

While dew and fog collection can provide a vital source of water in arid regions, the potential applications extend far beyond basic survival. Here are a few exciting possibilities:

Enhancing Agriculture: Captured dew and fog can be used to irrigate crops, particularly in areas where traditional methods are not sustainable. This can lead to increased food production and improved food security, especially in drought-prone regions.

Reforestation Efforts: Dew and fog collection can be used to support the growth of trees in dry areas. By providing a consistent source of moisture, these techniques can help to restore degraded ecosystems and combat desertification.

Emergency Water Supplies: Portable dew collectors and small-scale fog nets can be deployed in disaster situations where traditional water infrastructure is damaged or overwhelmed. This can provide a quick and reliable source of clean water for displaced communities.

Innovation and Sustainability

The future of dew and fog collection lies in continuous innovation and a focus on sustainability. Researchers are developing new materials and technologies to improve collection efficiency, making these techniques even more viable in diverse environments.

Additionally, integrating dew and fog harvesting with other sustainable water management practices, such as rainwater collection and greywater reuse, can create a comprehensive water security strategy for arid regions.

All in all, dew and fog collection offer a glimpse into a future where water scarcity is not an insurmountable obstacle.

By harnessing the power of condensation and embracing a more sustainable approach to water management, we can ensure a future where even the driest corners of the world can flourish.

This seemingly magical process, where the very air around us transforms into life-giving water, serves as a powerful reminder of nature's ingenuity and humanity's ability to adapt and innovate in the face of environmental challenges.

3.3: WINTER'S HIDDEN TREASURE: TRANSFORMING ICE AND SNOW INTO LIFE-GIVING WATER

Winter's arrival paints the world in a breathtaking blanket of white. While the beauty of snow and ice is undeniable, these frozen forms hold a hidden bounty – a vast reservoir of water. But unlike a rushing river or a sparkling lake, accessing this resource requires a different approach.

This section explores the art of transforming ice and snow into safe, drinkable water, a skill essential for survival in cold climates and a fascinating example of human ingenuity.

Melting Safely: Respecting the Chill

Extracting water from snow and ice may seem straightforward, but safety and efficiency are important. The first step is choosing the right source. Look for fresh, clean snow that's far from animal tracks, roads, or campsites. This minimizes the risk of contamination. When selecting ice, opt for clear, dense pieces; their compact structure suggests fewer trapped impurities.

Melting methods are crucial for preserving this precious resource. Our bodies naturally crave warmth, and instinct might lead us to gulp down snow directly. However, this can be dangerous, lowering our body temperature rapidly in an already cold environment.

Instead, gentle melting is key. Use the heat from a campfire or a portable stove to gradually transform the snow or ice into liquid. This conserves fuel and reduces the risk of introducing contaminants from burning materials.

A handy technique is the double-boiler method. Picture a pot of boiling water with a smaller container suspended inside. This inner container holds the snow or ice, allowing it to melt indirectly from the steam rising from the boiling water below. This prevents scorching and provides more controlled melting.

Maximizing the Winter Bounty

Planning ahead is crucial for making the most of winter's water reserves. Designate specific areas for snow and ice collection, keeping them away from potential sources of pollution. Think creatively - a sheltered area beneath trees or a shaded corner of your campsite can be ideal.

Nature itself can be your friend – burying snow or ice under a layer of fresh snow acts as a natural insulator, slowing down the melting process and keeping your water source frozen for longer.

The way you store snow and ice also matters. Compacting snow into solid blocks or storing large, unbroken pieces of ice minimizes their surface area exposed to the air. This slows down melting and extends your water supply. Remember, accessibility is important too.

Clear paths to your storage sites ensure you can easily reach your water source even as winter deepens its grip.

From Frozen to Drinkable: A Crucial Step

While snow and ice appear pristine, they can harbor unseen threats. Airborne pollutants can settle on these frozen surfaces, making purification essential before consumption. Filtration is the first line of defense. This can be as simple as straining melted snow or ice through a clean cloth, or using a more sophisticated portable filter.

But filtration alone might not be enough. Microscopic contaminants can still pose a risk. Purification methods address these threats. Boiling is the simplest and most effective method - bringing the water to a rolling boil for at least one minute eliminates most harmful bacteria and parasites.

For longer expeditions, consider chemical treatments with iodine or chlorine tablets. These readily available options disinfect the water, making it safe to drink. In scenarios with access to technology, ultraviolet light pens offer another method of purification.

Combining filtration with one of these purification techniques ensures your snow or ice-derived water is truly safe for human consumption.

Winter Survival Skills: Respecting the Elements

When snow and ice are your only water sources, survival hinges on knowledge and resourcefulness. Here are some valuable tips:

Pre-melting on the Go: If you're venturing into a cold environment, consider carrying a small container close to your body. The natural warmth radiating from your body can initiate the melting process for a small amount of snow or ice. This saves precious fuel for later and allows you to start melting more efficiently over a heat source when you reach camp.

Staying Warm While Melting: Hypothermia is a serious risk during winter survival situations. When melting snow or ice, keep a safe distance from the heat source to avoid sweating. Layer your clothing to regulate your body temperature and prevent chills.

If available, use reflective surfaces or shelters to trap heat near your melting container, further speeding up the process and minimizing your exposure to the cold.

Extracting water from snow and ice is a testament to the resilience of the human spirit. It's a process that demands respect for safety, efficiency, and the environment. Winter's seemingly harsh conditions transform into a valuable lesson in survival and sustainability.

Through careful planning, meticulous technique, and a deep appreciation for the natural world, the frozen stillness of winter reveals a hidden source of life, a testament to human adaptability and the enduring quest for survival.

3.4 UNEARTHING NATURE'S BOUNTY: A GUIDE TO TAPPING INTO UNDERGROUND WATER SOURCES

Just picture a vast network of hidden streams, not trickling on the surface, but coursing through the earth beneath our feet. These are aquifers, natural underground reservoirs that hold a significant portion of the planet's freshwater. In dry regions, these hidden veins are lifelines, providing a source of clean water for homes, farms, and communities.

But accessing this precious resource isn't as simple as sticking a shovel in the ground. It's a pursuit that blends science, experience, and respect for the land. Let's delve into the world of tapping into underground water sources, a practice that has evolved from its ancient roots of well-digging to embrace modern methods while staying true to principles of sustainability.

Finding the Hidden Gems: Signs and Science

The first step is detective work! Nature itself leaves clues about the presence of underground water. Look for the determined reach of plant life. Trees like willows and cottonwoods, known for their deep roots, often act as living indicators of water sources hidden below. Similarly, the very shape of the land tells a story. Depressions, dried-up riverbeds, all hint at the flow of water underground, guiding us towards potential wellsprings.

But these natural signs are just the beginning. Technology lends a helping hand with tools like ground-penetrating radar and seismic surveys. These sophisticated instruments act like X-rays for the earth, revealing the hidden landscape of water tables and rock formations.

However, despite these advancements, the process remains a beautiful interplay of knowledge and intuition. It's a conversation between the land and those who seek its bounty, ensuring a responsible and sustainable approach.

Building a Well: Planning and Safety

Once a potential water source is identified, meticulous planning comes into play. The choice between digging a well by hand or using a drill depends on the depth of the water table. Shallow aquifers, closer to the surface, can be accessed through manual excavation. Deeper reserves, however, require the precision and power of a drill.

Safety is paramount throughout this process. The type of terrain and the depth of the water table will determine the specific techniques used to prevent cave-ins and ensure the well's stability. Legal frameworks, which can vary by region, often dictate permits and well construction specifications.

Following these regulations is crucial, not just for legality, but for the well's long-term safety and the sustainability of the aquifer itself. We don't want to take more than the earth can replenish.

Bringing the Water Up: Pumps for Every Depth

Getting the water from the well to the surface requires some muscle, be it human or harnessed from the sun. Manual pumps are a reliable option for shallow wells, offering a simple and dependable way to extract water using nothing but human power.

For deeper sources, or for those seeking a more efficient solution, solar-powered pumps step in. These marvels of technology capture the sun's energy and use it to power the pump, highlighting the ever-evolving relationship between humans and the resources we rely on.

Choosing the right pump requires careful consideration. The depth of the well, the volume of water needed daily, and the available resources all play a part. Manual pumps become less effective with increasing depth, making powered alternatives more practical for deep aquifers.

Solar pumps, while having a higher initial cost, offer significant long-term benefits. They are environmentally friendly, free from the reliance on fossil fuels and electrical grids, and offer substantial savings over time.

Making Sure it's Safe to Drink: Water Quality Testing

Just because water comes from the earth doesn't mean it's automatically safe to drink. Contaminants, both natural and man-made, can find their way into even the most secluded aquifers, rendering the water unfit for consumption without proper treatment. Testing is essential to unveil these hidden threats - bacteria, heavy metals, and chemical pollutants.

Home testing kits provide a starting point, allowing you to detect common contaminants and get a general sense of your water's quality. For a more comprehensive analysis, professional laboratories can delve deeper, identifying the specific impurities present.

This detailed assessment, although requiring an investment, is crucial for safeguarding your health. By understanding the water's composition, you can implement targeted treatment solutions, ensuring your well serves as a source of life-giving water, not illness.

A Pact with the Earth: Responsibility and Sustainability

Tapping into underground water sources is a venture into the very heart of the earth. It's a pact with the land that demands respect, knowledge, and a deep sense of responsibility.

The entire process, from locating hidden reserves to ensuring the purity of the water, underscores the interconnectedness of humanity and the natural world. It's a constant reminder that the resources we depend on are not ours for the taking, but require careful stewardship.

3.5 ENSURING LONG-TERM SUSTAINABILITY: PROTECTING OUR VITAL RESOURCE

Understanding the limitations of aquifers is key to maintaining their health and ensuring they continue to provide water for generations to come. Here's how we can be responsible stewards:

Sustainable Withdrawal: Just like a bank account, aquifers have a balance. We can only withdraw (pump out) as much water as the aquifer can naturally replenish (through rainfall and snowmelt). Overexploitation, taking out more water than is replaced, can lead to a lowering of the water table, reduced well yields, and even land subsidence (sinking of the ground).

Monitoring Water Levels: Regularly monitoring water levels in your well is crucial. A consistent decrease can be a sign of overexploitation and a wake-up call to reduce water usage or explore alternative sources.

Water Conservation Practices: Every drop counts! Implementing water-saving practices at home and on the farm goes a long way.

This includes fixing leaky faucets, using low-flow showerheads, and adopting efficient irrigation techniques.

Protecting Water Quality: Activities on the land can impact the quality of groundwater. Proper disposal of waste, avoiding the use of harmful pesticides and fertilizers near wellheads, and creating buffer zones around wells are all essential steps in safeguarding the purity of our water source.

Community and Collaboration

Groundwater isn't an individual resource; it's a shared treasure. Working together as a community is vital for long-term sustainability. Here are some ways communities can collaborate:

- Data Sharing: Sharing well data, water usage patterns, and aquifer health information fosters a collective understanding of the resource and empowers communities to make informed decisions about water management.
- Joint Management Plans: Developing and implementing joint management plans that outline responsible withdrawal practices, conservation goals, and pollution prevention strategies ensures a unified approach to protecting the aquifer.
- Investing in Education: Spreading awareness about the importance of groundwater, its vulnerability, and sustainable water use practices empowers individuals to become active participants in protecting this vital resource.

By embracing these principles of responsibility, collaboration, and continuous learning, we can ensure that underground water sources continue to be a source of life for generations to come.

CHAPTER 4
STOCKPILING YOUR DRINKING WATER: A SIMPLE GUIDE FOR HOME STORAGE

S uppose a power outage or a bad storm hits your area. Having a safe supply of water at home is like having a secret weapon! This chapter will show you how to turn regular tap water into a reliable resource you can count on. We'll cover what containers to pick, how much water you'll need, and the super important step of keeping everything clean.

4.1 PICKING THE PERFECT CONTAINER: FIND THE RIGHT FIT FOR YOU

The container you choose for your water storage is kind of like picking out a new backpack. It should be the right size and type for your needs. Here are some popular options:

Plastic drums: These are common and lightweight, making them easy to move around. They're kind of like big water jugs, but they can hold a lot more water. However, some plastics can release harmful chemicals, so be sure to look for ones that say "BPA-free" on the label. BPA can make you sick, so it's best to avoid it.

Stainless steel tanks: These are super strong and won't leak chemicals, making them a safe choice for your water. They're like giant thermoses, but they're not insulated and can be expensive and heavy. If you have limited space or aren't looking to store a ton of water, these might not be the best option for you.

Glass bottles: These are another safe option because glass doesn't release any chemicals. You can also see right through the glass to check how clean the water is. However, glass bottles can break easily, so you need to be careful with them. They might not be the best choice if you have small children or pets running around.

Safety First!

This might seem obvious, but it's super important! You don't want anything yucky contaminating your water and making you sick. Here are some safety tips to remember:

- Never reuse containers that hold things like soap or cleaning supplies. Leftover residue from these products can get in your water and make you sick.
- Choose BPA-free plastic to avoid health risks. BPA is a chemical that can mess with your hormones and might not be good for you in the long run.

How Much Water Do You Need? Planning for Every Sip

Imagine yourself making a grocery list. You need to figure out how much water you use every day for things like drinking, cooking, and cleaning. Then think about how much extra water you might need in case of an emergency, like a bad storm or a broken pipe.

The checklist at the end of this section will help you figure out the perfect amount for your family.

Getting Your Containers Ready: Cleaning Up for Fresh Water

Before you fill them with water, you need to give your containers a good cleaning. This removes any leftover manufacturing materials or dirt that could make you sick. Think of it like washing your dishes before you use them! Here's how to clean each type of container:

For plastic and metal: Wash them with a solution of bleach and water, then rinse them really well. You want to make sure all the bleach is gone before you add any water.

For glass: Boil the bottles in clean water on the stove. This will kill any bacteria that might be hiding on the glass.

This ensures your water stays clean and safe to drink, just like you like it!

Here's a Handy Checklist to Keep Your Water Safe:

- Container History: Only use new containers or ones previously used for food.
- Material Safety: Pick BPA-free plastic, stainless steel, or glass to avoid chemicals leaching into your water.
- Water Amount: Figure out how much water you need based on daily use and emergencies (consider the number of people and pets in your household).
- Space Check: Measure your storage area and consider factors like temperature and sunlight. Extreme heat or cold can affect the quality of your water.
- Cleaning Up: Clean and sterilize containers before use (use the right method for each material).

- Regular Inspections: Check your containers for cleanliness and damage regularly, especially after extreme weather events. A broken container could lead to water contamination.

4.2 WHY WE NEED TO ROTATE OUR WATER STORAGE

Storing water for emergencies is a smart move, but just leaving it untouched isn't enough. Over time, even perfectly safe water can become unclean or develop an unpleasant taste. That's why we need to rotate our water supply, which means using the older water and replacing it with fresh water.

Here's the science behind why water rotation is important:

Germ Growth: Stagnant water, just like a pond that never gets refreshed, can become a breeding ground for tiny organisms we can't see, like bacteria and algae. These little guys aren't good for us, and we definitely don't want them in our drinking water!

Container Breakdown: Even the best containers we choose for storing water aren't perfect. Over a long period, even these containers can slowly release small amounts of chemicals into the water, affecting its taste. It's kind of like how food stored in plastic containers for too long can sometimes taste a bit off.

How Often to Rotate Water:

A good rule of thumb for rotating your water supply is every six months, around the spring and fall equinoxes. These equinoxes are like natural markers in the year, making it easier to remember to check your water. But there are some situations where you might need to rotate your water more often:

- Hot Climates: If you live in a place that gets very hot, the heat can speed up the breakdown of containers and the growth of germs in your water. In this case, consider rotating your water every three months instead.
- Direct Sunlight: Sunlight can also have a negative impact on stored water. If your water containers are exposed to direct sunlight, you might want to rotate them every three to four months to be safe.

Keeping Track of Your Water

Keeping track of your water storage is important to make sure you use the older water first. Here are some simple tips:

Labeling is Key: Grab a marker and write the date you filled each container on the outside. This way, you can easily see which water is the oldest and needs to be used first.

First In, First Out: Imagine your water storage is like a grocery store line. The containers you put in first (the oldest ones) should be the ones you use first. To make this happen, store newer containers in the back or on the bottom of your storage area. This is a fancy way of saying "first in, first out," and it's a great way to ensure you're always using the oldest water.

Making Stored Water Taste Better

Even if you've been diligently rotating your water, it might not always taste as fresh as spring water. Here are a couple of tricks to improve the taste of stored water:

Get Some Air In There: Water that's been sitting still for a while can lose some of its oxygen, which can make it taste flat. To add some oxygen back in, simply pour the water back and forth between two containers a few times. This process, called aeration, is like taking a deep breath for your water!

Filter It Out: Before drinking your stored water, consider running it through a filter. This will remove any particles that might have formed in the water while it was in storage, making it clearer and more pleasant to drink.

Using and Replacing Your Water

The best way to keep your water fresh is to actually use it! Here's a clever way to integrate using your stored water into your daily routine:

Make it Part of Your Routine: Set aside some of your stored water for everyday tasks like cooking, cleaning, or watering your plants. This way, you're constantly cycling through your older water supply and replacing it with fresh water. It's like using a refillable water bottle – you're constantly using the water and refilling it, keeping everything fresh!

Replenishing Your Water Supply

When you replace the water you've used, take a moment to give your storage area some love:

Container Check-Up: Just like you check your car tires for wear and tear, take a moment to inspect your water containers for any damage or leaks. Damaged containers can compromise the quality of your water, so it's important to make sure they're in good shape.

Cleaning Up: Before refilling your containers, give them a good cleaning with soap and warm water. This will remove any dust or dirt that might have accumulated

4.3 BUG-OUT WATER CACHES: A GUIDE FOR BEGINNERS

This guide explains how to hide water along your escape route in case of an emergency, like a fire, flood, or another disaster. Having hidden water supplies can be very important for survival.

Suppose you are on the road, away from home because of a big event. If you have water hidden along the way, you won't have to worry about running out of something so essential for staying alive.

Our bodies are mostly water, and we need it to function properly. Even mild dehydration can cause headaches, tiredness, and dizziness. In an emergency situation, these symptoms could be dangerous. Having hidden water caches can give you peace of mind, knowing you have a reliable source of hydration on your journey.

Choosing Hiding Spots:

Pick places that are easy for you to get to when you need them, but not easy for others to find. Think about it this way: if you're running from danger, you don't want to waste time searching for your water. But you also don't want someone else to stumble upon it and take it for themselves.

Imagine yourself following your escape route under pressure. You should be able to find your water caches quickly and easily, without getting lost or confused. Look for places that are recognizable but not obvious. For example, a large, hollowed-out log near a stream might be a good spot.

The stream itself is a landmark that will help you remember the location, but the specific log might not be immediately noticeable to someone else.

Think about the landscape: Hills, rivers, big rocks, or even special trees can help you find your stash later. Use landmarks as clues to remember where you hid your water. The big oak tree at the bend in the road, the wide stream you have to cross - these can all be reminders. Let's say you're hiding water near a steep hill. You could make a note on your map that says "Water hidden behind the rocky outcrop, halfway up Smith's Hill." This description will help you pinpoint the exact location later.

Try to hide your water off the main path you'll be taking. If everyone is following the same escape route, it's best to hide your water a little bit away from the main trail. This way, it's less likely to be found by others. Imagine a narrow forest path. Instead of hiding your water right next to the path, you could venture 10 feet off the trail and tuck it beneath a cluster of bushes. This small detour will add a layer of security to your water cache.

Keeping the Water Safe:

Use containers that won't leak or break. A leaky container will spill all that precious water when you need it most! Plastic jugs or canteens are good options, but make sure they are sealed tightly. You can also find containers specifically designed for long-term water storage. These containers are usually made from tough, BPA-free plastic and have airtight lids to prevent leaks and conta-mination.

Wrap them in plastic bags for extra protection. Just like an extra layer of clothing keeps you warm, an extra layer of plastic keeps your water safe from dirt, dust, and anything else that might try to get in. Look for heavy-duty plastic bags that won't tear easily. You can also use a combination of a plastic bag and a waterproof container for ultimate protection.

You can also buy special containers made for storing water outside. These containers are designed to be tough and water-proof, so your water will stay fresh and clean no matter what nature throws at it. Look for containers that are UV-protected to block sunlight, which can degrade the quality of the water over time.

Pick spots away from animals:

Steer clear from spots that might be curious or thirsty and might try to get to your water. We all love animals, but a thirsty raccoon can be a real problem! Try to hide your water away from areas where animals like to wander and look for food. Smell-proof bags can also help here. Consider areas with low animal traffic, or places where the natural landscape discourages them, like areas with lots of rocks or dense undergrowth.

4.4 ADVANCED WATER STORAGE: BEYOND THE BUCKET

This section dives into sophisticated methods for long-term water storage, ditching the simple bucket for high-tech solutions. These systems go beyond just storing water – they protect it and can even benefit the environment.

Underground Water Chambers:

Forget buckets – imagine a whole room built underground specifi-cally for water! These underground chambers, called cisterns, are dark and cool, perfect for keeping water fresh. Sunlight and conta-mination can't reach it there. The materials used to build these tanks are super strong and won't leak, allowing them to hold large amounts of water.

Turning Rainwater into Tap Water:

This system creates a closed loop for water. Rain falling on your roof gets collected, treated, and then used for toilets, washing machines, or gardens. It's like a miniature version of nature's water cycle, reducing your reliance on city water. This is good for your wallet and the environment!

Keeping the Water Safe:

Even the best systems need protection from unwanted guests in the water, like germs or insects. Some systems use natural filters made of sand, gravel, and charcoal, mimicking how the earth itself purifies water. Tight-fitting lids are also crucial for keeping everything out.

The Future of Water Storage:

New inventions are making water storage even better. Special containers made with algae-resistant materials prevent your water from getting scummy, keeping it clean for longer. Smart monitors act like little water watchdogs, checking water quality, level, and temperature. They can even warn you of problems before they become major issues. This fancy tech ensures your water stays safe and usable.

More Than Just Storing Water:

By using these smart water storage methods, we're doing more than just preparing for emergencies. We're showing respect for our planet by using less water overall, relying more on rainwater, and keeping the water we do have clean. It's all about coexisting with nature.

From Underground Vaults to Smart Systems:

So, whether it's hiding water underground, capturing rainwater, or using the latest technology, every step we take to store water helps us be more prepared. It's a testament to human ingenuity and adaptability. By following these ideas, we can improve not only how we store water but also how we see our place in the world.

CHAPTER 5

MAKING DIRTY WATER SAFE TO DRINK

Clean water is what keeps us alive, just like air. But sometimes, the water we find in lakes, streams, or even at home might not be safe to drink. It could have tiny living things in it that can make us sick, or even invisible chemicals. This chapter will explain different ways to make dirty water clean and safe to drink.

5.1 BOILING, FILTERING, AND CHEMICALS: GOOD AND BAD SIDES

There are three main ways to clean dirty water: boiling, filtering, and using chemicals. Each way has its pros and cons, so let's take a look at them one by one.

Boiling:

This is a very old way to clean water, and it's still a good choice today. Here's why:

- Good: Boiling is great at killing germs. It gets rid of bacteria, viruses, and even nasty little parasites that can make you very sick. It's also a simple way to clean water – you just need a pot and a fire.
- Bad: Boiling won't remove chemicals from the water. If there are things like pesticides or other pollutants in the water, boiling won't get rid of them. Also, you need a fire to boil water, which can be a problem if you're out camping or there's a power outage.

Filtering:

Filters are like sieves for water. They let the water pass through, but they trap bigger things like dirt, sand, and even some germs. Here's what you need to know about filtering:

- Good: Filtering is a good way to get rid of dirt and some germs that can make you sick. You can even make a simple filter yourself using cloth or a coffee filter!
- Bad: Not all filters are created equal. A simple cloth filter won't stop the tiniest germs, like viruses, from getting through. Better filters that can stop these tiny germs are more expensive and need to be cleaned regularly so they keep working well.

Chemicals (like iodine or chlorine):

Chemicals like iodine tablets or chlorine drops can also be used to clean water. Here's what you should know about them:

- Good: Chemicals can kill germs very quickly and easily. They're also portable, so you can take them with you wherever you go.

- Bad: Chemicals can leave a bad taste in the water, which some people might not like. Also, chemicals like iodine aren't good for everyone, especially pregnant women. It's important to check with a doctor before using iodine to clean water.

Working Together for the Best Results

As you can see, each way of cleaning water has its ups and downs. No single method is perfect. The best way to clean water often involves using more than one method at the same time.

Here's an example: Imagine you find a stream in the woods. The water looks dirty and there might be farms nearby that use pesticides. In this case, you could use a filter first to remove the dirt and some germs. Then, you could boil the water or add a chemical like iodine to kill any remaining germs that the filter might have missed.

A Handy Checklist to Help You Choose

Choosing the best way to clean your water can seem tricky, but don't worry! This checklist will help you out:

1. Take a good look at the water: Is it cloudy or dirty? Are there factories or farms nearby that might be polluting the water?
2. Pick a main cleaning method: Based on what you saw in step 1, decide if boiling, filtering, or chemicals would be the best first step.
3. Don't forget a second method: Remember, no single method is perfect. Choose a second method to get rid of anything the first method might have missed.

4. Taste test: After you've cleaned the water, give it a taste. If it tastes funny or unpleasant, you might need to boil or filter it again.

5. Clean your filter: If you're using a filter, make sure to clean it regularly according to the instructions. This will keep it working well and remove germs from your water.

Using these steps, you can turn questionable water into safe drinking water! Know that clean water is essential for good health, so don't be afraid to take some time to make sure the water you drink is safe.

5.2 MAKING YOUR OWN WATER FILTER AT HOME (SUPER EASY!)

When would this be helpful?

Imagine you're on a camping trip or maybe the faucet water at home isn't tasting so good. Maybe it looks cloudy or smells funny. That's where these special filters come in! They can help turn icky water into clean water that's safe to drink. And the best part? You can make them yourself with things you might already have around the house!

What you'll need for a charcoal filter:

Charcoal: This is like a black sponge for yucky stuff in the water. You can use charcoal from burnt wood, but make sure it's from hardwood like oak or birch, not pine or other softwoods. Softwood charcoal might not be good for you.

Sand: This helps trap tiny dirt particles that the charcoal might miss. Play sand from the beach or sandbox won't work because it's too fine. Look for sand that feels gritty between your fingers, like the kind in a fish tank.

Rocks (different sizes): Bigger rocks on the bottom help water flow through easily, and smaller rocks on top help trap even more dirt.

How to build a charcoal filter:

1. Grab your container: This could be a big bucket, a pot with a wide opening, or anything else that can hold water and has two open ends (a top and a bottom).
2. Rock bottom: Put a layer of bigger rocks in the bottom of your container. This will act like a little highway for the water to flow through easily.
3. Sandcastle layer: Next, add a layer of sand on top of the rocks. This layer will catch any small dirt particles that the rocks let through.
4. Charcoal magic: Now comes the superstar of the filter - the charcoal! Crush the charcoal into pieces that are about the size of peas. Then, add a good layer of this crushed charcoal on top of the sand. Remember, the charcoal is like a sponge, soaking up all the bad things in the water!
5. Pour and wait: Now that your filter is built, carefully pour your dirty water into the container. As the water trickles down through the rocks, sand, and charcoal, it gets cleaned up along the way! The clean water will collect at the bottom of the container.

What you'll need for a bottle filter (perfect for on-the-go!):

- A plastic bottle: This will be your mini water purifier!

- Scissors: You'll need these to cut the bottle. Make sure an adult helps you with this part!
- Stuff to filter the water: Same as the charcoal filter, you'll need charcoal (in small pieces), sand, rocks, and cloth.

How to build a bottle filter:

1. Cut the bottle in half: Ask an adult to help you cut the bottom part of the bottle off with scissors.
2. Cloth catcher: Cut a small piece of cloth, like a bandana or a clean coffee filter, to fit snugly inside the opening of the bottle top. This will stop big chunks of dirt from getting through.
3. Charcoal power: Add a layer of your pea-sized charcoal on top of the cloth.
4. Sand sensation: Now add a layer of sand on top of the charcoal.
5. Rock on!: Finally, add a layer of small rocks on top of the sand.
6. More layers for cleaner water (optional): If your bottle is big enough, you can add more layers of cloth, sand, and rocks for even better filtering. The more layers, the cleaner the water will be!
7. Pour and drink!: Carefully pour your dirty water into the open top of the bottle. The water will seep down through the filter layers, getting cleaner with each layer it passes through. The clean water will drip out of the bottle cap, ready for you to drink!

Keeping your filter clean:

These filters are like little sponges, and sponges get dirty after a while! The charcoal and sand will eventually get clogged up with all the yucky stuff they take out of the water. Here's what to do to keep your filter working well:

- Throw away the old stuff: Every few times you use your filter, you'll need to throw away the old charcoal and sand. They'll be full of the bad stuff they filtered out of the water.
- Replace with fresh: After throwing away the old stuff, refill your filter with new charcoal and sand.
- Rock washing: The rocks in your filter can also get dirty. Every now and then, take them out and rinse them with clean water to get rid of any built-up gunk.

Important!

These DIY filters are lifesavers in a pinch, but it's important to know they don't zap every single bad thing out of the water. They can't get rid of very tiny things like viruses or some chemicals that might make you sick. Here's what to keep in mind:

- Boil for extra safety: If you're ever worried that your water source might be really dirty, it's always best to boil the water for at least a minute after filtering it. Boiling will kill any nasty little critters that the filter might have missed.
- Get it tested (optional): If you're unsure about your water quality, especially if you're using water from a new source like a lake or stream, you can take a sample to a water testing facility. They can check for things like bacteria and chemicals that your filter can't handle.

5.3 HARNESSING THE SUN'S POWER: SOLAR WATER DISINFECTION (SODIS)

In a world where complicated solutions often take center stage, a method called Solar Water Disinfection (SODIS) stands out for its simplicity. It uses the sun's natural power to make dirty water safe to drink.

Here's how it works:

SODIS relies on two things: sunlight and heat. The sun's rays, especially the ultraviolet (UV) light, kill harmful germs in the water, making them unable to spread sickness. At the same time, heat from the sun (over 113°F or 45°C) helps kill even more germs. This combination of sunlight and heat makes SODIS a powerful tool for purifying water, especially in places where clean water is scarce.

Using SODIS is very easy:

1. Grab clear plastic bottles: Look for bottles made of PET (polyethylene terephthalate). These are the most common clear plastic bottles used for drinks. Make sure they are clean and free of any stickers or labels.
2. Fill the bottles with water: If the water is cloudy, try straining it through a clean cloth to remove any dirt or particles.
3. Let the sun do its magic! Lay the bottles flat on a sunny surface, like a roof or a table. The more sun exposure, the better. On a clear, sunny day, six hours of sunlight are enough. If it's cloudy, you might need to leave the bottles out for two days.

Why is SODIS so great?

- It's practically free: All you need are some clear bottles and sunshine, which is free and available almost everywhere.
- It's easy to use: You don't need any fancy equipment or special skills to use SODIS.
- It's effective: Studies have shown that SODIS can kill a large number of germs in water.

However, SODIS does have some limitations:

- It can't remove chemicals: SODIS won't remove harmful chemicals like pesticides or heavy metals from the water.
- It works best with clear water: If the water is very cloudy or dirty, SODIS might not be effective enough.

Making SODIS even better:

There are some ways to improve SODIS and make it work even better:

- Use reflectors: Placing mirrors or other reflective surfaces around the bottles can increase the amount of sunlight they receive and shorten the disinfection time.
- Use temperature indicators: These are small stickers that change color when the water reaches a safe temperature, giving you peace of mind that the germs are gone.

How to know your water is safe:

After using SODIS, it's a good idea to check the water to make sure it looks, smells, and tastes good. You can also use special test kits to check for the presence of germs. This double-check helps ensure the water is truly safe to drink.

SODIS is a powerful example of how we can use nature's forces to our advantage. By harnessing the power of the sun, we can turn dirty water into clean drinking water, a simple yet vital tool for survival, especially in areas with limited resources.

5.4 ADVANCED FILTRATION SYSTEMS: HIGH-TECH SOLUTIONS FOR CLEAN WATER

When your water worries go beyond basic dirt and germs, advanced filtration systems come to the rescue. These are like high-tech bodyguards for your water, protecting it from tiny threats and hidden dangers.

There are several options available, each tackling different problems:

Reverse Osmosis (RO): This is the ultimate purifier, using a special filter to remove almost everything bad in your water, including bacteria, viruses, and even dissolved chemicals like lead. It's like a super-sieve that only lets water molecules through. However, RO systems can be expensive and require electricity to run.

Ultraviolet (UV) Purification: This method uses ultraviolet light, like what's in a tanning bed but much stronger, to zap bacteria and viruses, making them harmless. It doesn't change the water's taste or remove chemicals, but it's a good option for killing germs without any fuss.

Advanced Carbon Filters: These filters are like tiny sponges with a huge surface area. They trap organic chemicals like pesticides and chlorine, improving the taste and smell of your water while leaving the good minerals behind. They're generally affordable and easy to maintain.

Choosing the right system depends on your needs and budget:

- For the Best Purification: Reverse osmosis is king, but it comes with a price tag and needs electricity.
- For Fighting Germs on the Go: Portable UV purifiers are lightweight and battery-powered, perfect for disinfecting water on camping trips.
- For Improved Taste and Odor: Advanced carbon filters are a good first line of defense, especially if you're concerned about taste or specific chemicals.

Home and Away, Staying Protected:

At home, you can install a whole-house filtration system with multiple stages, each tackling different contaminants. This way, every drop coming out of your tap is guaranteed to be clean. While this requires an upfront investment, it gives you peace of mind, especially if you're worried about your local water quality.

For those who are always on the move, there are portable options too. Reverse osmosis and UV purifiers come in travel-friendly versions, ensuring clean water wherever you go.

Remember, all filtration systems need regular care. Filters and membranes get clogged over time, so cleaning and replacing them is crucial to keep your system working effectively. It's a small price to pay for the constant flow of pure water.

Advanced filtration systems are a testament to human ingenuity. They're not just tools, they're a promise of clean water, even when the world throws challenges our way. By choosing the right system and taking care of it, you can be sure that every drop you drink, at home or on the go, is safe and refreshing.

CHAPTER 6
THE INVISIBLE ADVERSARIES: CHEMICAL POLLUTANTS IN WATER

I magine a glass of water. It looks clean, but a silent war is happening within it. Tiny, harmful substances we can't see are polluting our water. These are chemical pollutants, and they come from many sources of modern life. This chapter will explain what these pollutants are, how to find them, and how to get them out of our water.

6.1 DIFFERENT TYPES OF POLLUTERS

- Farms: Farmers use pesticides and weed killers to protect their crops. When it rains, these chemicals can wash off the fields and into rivers, lakes, and underground water sources.
- Factories: Industries release a mix of chemicals during production. These can include heavy metals, which are very toxic, and solvents used for cleaning.

- Homes: Even everyday products like medicine and cleaning supplies can pollute water. When we throw them away, they might end up in sewage systems that eventually reach rivers or oceans.

These pollutants can make us sick, both in the short term (like causing a stomach ache) and over a long period of time (like increasing the risk of cancer). They can also harm the plants and animals that live in water.

Finding the Culprits

There are ways to check for these invisible enemies. Simple test kits can be used at home to see if there are common contaminants in your water. However, for a more detailed picture, professional lab testing is needed. This testing can identify the specific types of pollutants and how much of them are present.

Cleaning Up Our Water

Once we know what's in the water, we can take steps to clean it. Here are a few methods:

Activated Carbon Filters: These filters have tiny holes that trap organic chemicals and chlorine, making the water taste better.

Distillation: This method involves boiling water and collecting the steam as it condenses back into liquid. This removes most contaminants, except those that boil at a higher temperature than water.

Chemical Neutralization: This method uses specific chemicals to change harmful pollutants into harmless substances. It's like a targeted attack on the bad guys in the water!

Choosing the best method depends on what kind of pollutants are in your water.

Be Prepared: Your Water Safety Plan

What if your water supply gets contaminated? It's important to have a plan in place. This plan should include:

- Backup Water Source: Find another source of clean water you can use in case your main supply is polluted. This could be bottled water or a spring nearby.
- Treatment Options: Based on the type of contamination, know which purification method to use to make your water safe again.
- Regular Updates: Keep your plan up-to-date as things change or new information becomes available.

Checklist for Safe Water

Here's a handy checklist to help you fight against chemical pollutants in your water:

1. Identify Possible Sources: Think about nearby farms, factories, or cities that could be polluting your water.
2. Choose a Detection Method: Decide between a home test kit for a quick check or professional lab testing for a more detailed analysis.
3. Identify the Pollutants: Based on the testing, find out what specific chemicals are in your water.
4. Pick a Cleaning Method: Choose the best water purification method (filtration, distillation, or neutralization) for the pollutants you found.
5. Consider Alternatives: If your current treatment isn't enough, explore options like using bottled water or additional purification methods.

6. Review and Update: Regularly check if your plan is working and update it based on new information or changes in your water quality.

This is how you can take charge of your water safety and ensure you have clean, healthy water for drinking and everyday use.

6.2: BIOLOGICAL THREATS IN WATER: VIRUSES, BACTERIA, AND PROTOZOA

While water sustains life, hidden within its depths lies a potential threat: biological contamination. Microscopic organisms like viruses, bacteria, and protozoa can transform this essential resource into a breeding ground for disease.

Invisible Enemies: Types of Biological Threats

- Viruses: These minute pathogens, the smallest of the three, infiltrate water molecules and pose a significant health risk. Examples include norovirus and hepatitis A, capable of surviving in water for extended periods. Their presence often indicates fecal contamination, highlighting the importance of proper sanitation for safe water.
- Bacteria: Slightly larger but equally concerning are bacteria like E. coli and Vibrio cholerae. Their growth in water frequently stems from similar sources of contamination, underlining the consequences of inadequate waste management.
- Protozoa: Completing the triad of biological threats are protozoa like Giardia and Cryptosporidium. Encased in protective cysts, they exhibit remarkable resilience in

water, persisting through conditions that would eliminate their viral and bacterial counterparts.

The health risks associated with consuming water contaminated by these microorganisms range from mild to severe. While initial responses may include diarrhea, vomiting, and fever, the impact can extend beyond acute symptoms, potentially leading to chronic health problems.

Combating the Threat: Purification Methods

Effective water purification methods are the cornerstone of our defense against biological contaminants.

Boiling: This simple yet universal solution effectively neutralizes viruses, bacteria, and protozoa. Bringing water to a rolling boil disrupts the very structure of these microorganisms, rendering them harmless.

Ultraviolet (UV) Treatment: A more modern approach, UV treatment utilizes ultraviolet light to penetrate microbial cells, damaging their DNA and preventing them from causing harm. While efficient and rapid, this method requires access to UV technology and clear water for optimal performance.

Filtration Technologies: Microfiltration and ultrafiltration systems target the physical removal of pathogens. Equipped with filters of specific pore sizes, these systems act as barriers, capturing microorganisms while allowing clean water to pass through. Each of these methods plays a vital role in securing safe drinking water.

Prevention: The Best Defense

The most sustainable solution to biological contamination lies in preventative measures. Proper waste disposal minimizes the risk of fecal matter entering water sources. Establishing buffer zones around water sources and treating sewage before release creates a barrier that shields water from contamination.

Educating communities about hygiene, the risks of untreated water, and available purification methods fosters a culture of vigilance. By implementing these measures collectively, we not only address the immediate threat of biological contaminants but also contribute to broader public health and environmental stewardship.

6.3: HEAVY METAL MENACE: THREATS AND SOLUTIONS FOR CLEAN WATER

Water is essential for life, but hidden within its depths can be a silent threat: heavy metals. These toxic elements, like lead, arsenic, and mercury, seep into our water from various sources, both old and new. Industrial waste, farming practices using metal-based pesticides, and even aging pipes can all contribute to this invisible contamination.

The dangers of heavy metals aren't always immediate. They can build up in your body over time, causing health problems that may not show up right away. Lead, for example, can damage the nervous system, especially in children, affecting their learning and development.

Arsenic is another sneaky culprit. It can slowly increase your risk of getting cancer, heart disease, and diabetes. Mercury, though less common, can harm your brain and kidneys.

Because the effects can be subtle and people react differently to these metals, it's important to be proactive about keeping your water clean. Some people, like babies, pregnant women, and those already sick, are more at risk from heavy metals, so ensuring clean water for everyone is crucial.

Fighting Back: Ways to Remove Heavy Metals

There are technological solutions available to remove heavy metals from water, but the best method depends on the specific metal you're dealing with.

- Reverse Osmosis Systems: These systems use special filters to remove almost all lead, arsenic, and mercury. However, they require regular maintenance and may not be affordable for everyone.
- Ion Exchange: This method is like a swap meet for atoms. It replaces harmful metal ions in your water with safe ones, making the water less toxic. The type of resin used depends on the specific metal you want to remove.

Not everyone has access to these advanced systems, which is why regular water testing is so important. Testing reveals the presence of heavy metals and helps you decide how to address them. Based on the test results, you can choose the best treatment option or explore alternative water sources.

Taking Charge of Your Water Safety

If you're worried about heavy metals, there are steps you can take to protect yourself and your family:

Upgrade Your Water Treatment: Consider installing better filters or adding specialized treatment methods to your existing system. This is an investment in your health.

- Explore Alternative Sources: Collect rainwater and purify it carefully, or tap into a protected well if available. Remember, each alternative source has its own quality and sustainability considerations.

When you comprehend the health risks of heavy metals and learn about the different removal methods, you become empowered to take action. Regular testing helps you stay ahead, and implementing the right solutions protects you from these invisible threats. This proactive approach ensures your water remains a source of life, not illness. It's about more than just survival; it's about thriving in the face of environmental challenges.

6.4: NATURE'S WATER PURIFIERS: HARNESSING EARTH'S CLEANSING POWER

Where land and water meet, a remarkable phenomenon unfolds. Natural filtration systems, like wetlands and sand filters, silently transform murky water into clean, usable sources. These systems rely on a combination of processes – adsorption, sedimentation, and biological degradation – to remove contaminants. Each step leverages the natural properties of Earth's materials, working in concert to purify water within the ecosystem.

Sand Filtration: A Simple but Effective Method

Sand filters employ layers of sand and gravel as a medium for water to pass through. As water percolates through these layers, larger particles get trapped by the granular texture of the sand. This mimics how rain naturally filters through the earth, removing impurities. Beyond simple physical trapping, adsorption also occurs on the surface of the sand grains. Organic compounds cling to the sand, preventing them from moving further.

Sedimentation: Letting Gravity Do the Work

Sedimentation further aids in removing suspended particles. The reduced water flow through the filter allows gravity to pull these particles down, settling them away from the clean water that eventually emerges.

Wetlands: Nature's Complex Purification System

Wetlands are a prime example of natural filtration at its most intricate. Here, water meanders through a complex matrix of plant roots, soil, and microbial communities. Each component plays a vital role in degrading and removing pollutants.

Plants act as natural filters, absorbing nutrients and even some heavy metals into their biomass. Microbes break down organic compounds, rendering them harmless. The soil itself functions as both a physical filter and a chemical reactor. Processes like denitrification transform pollutants into harmless gasses that are released into the atmosphere.

DIY Natural Filtration for Home Use

The beauty of these natural filtration systems lies in their accessibility. Individuals can harness these processes on a smaller scale to create effective water purifiers for home use. A simple sand and gravel filter can be assembled with readily available materials. Layers of sand and gravel placed in a container allow water to be poured through and collected after filtration. These systems require minimal maintenance – primarily the occasional replacement of the sand to ensure ongoing effectiveness.

For those with more space, constructed wetlands can be integrated into landscaping. These small-scale wetlands mimic their natural counterparts, with specific plants and carefully designed water flow paths to maximize their purifying capabilities. They not only provide water treatment but also create habitat for wildlife.

Combining Natural and Mechanical Methods for Optimal Results

While natural filtration systems are powerful, they have limitations. Their ability to remove some contaminants, particularly specific pathogens and heavy metals, can be limited. A thoughtful integration of natural systems with mechanical purification methods is necessary for comprehensive water treatment solutions.

For instance, water treated in a constructed wetland could then be passed through a UV purifier to eliminate any remaining pathogens. Similarly, water filtered through a sand and gravel system could be further purified by reverse osmosis to remove dissolved chemicals. This multi-barrier approach leverages the strengths of each method to achieve a higher level of water purity.

A Sustainable Approach to Clean Water

The synergy between natural and mechanical systems goes beyond just water purity. It underscores a broader philosophy of sustainability. These natural filtration systems operate without the need for chemical additives or significant energy inputs. Their maintenance demands are low, and they contribute positively to the local ecosystem.

Moving Forward: A Collaborative Approach to Water Security

By embracing these integrative approaches, individuals and communities can secure cleaner water while fostering ecological balance. This paves the way for a future where the health of our waterways and the well-being of our planet are held in equal regard.

This exploration of natural filtration highlights the potential of Earth's resources and the importance of working with nature, not against it. By combining natural and mechanical methods, we can ensure clean water for ourselves and future generations, all while protecting the delicate balance of our environment.

CHAPTER 7

DIY WATER PURIFICATION: TAKING CHARGE OF YOUR WATER

I magine being able to turn any water source into something safe to drink using just a few basic tools. That's the power of DIY water purification, a valuable skill for anyone who wants to be self-sufficient. In this chapter, we'll focus on distillation, a process that's like a rite of passage for those who want to be independent from relying on outside sources for clean water. We'll explore how to build a simple water still and turn it into a tool for survival.

7.1 BUILDING YOUR OWN WATER PURIFIER: UNDERSTANDING DISTILLATION

Distillation is based on a natural process as old as clouds themselves. Think about how rain forms: water evaporates, leaving impurities behind, then condenses back into pure rain. Water still mimics this cycle.

By heating water, we turn it into vapor, which then cools and condenses back into a liquid state, leaving most contaminants behind because they can't vaporize at the same temperature as water. This simple yet effective method allows you to get clean drinking water from sources that may be salty, filled with bacteria, or contain dissolved chemicals.

Building Your Still: What You'll Need

The good news is that you can build a basic water still with things you might already have at home or find outdoors. Here's what you'll need:

- A pot for boiling water
- A smaller container to collect the purified water
- A lid that curves inward (convex) or a sheet of plastic wrap
- A heat source (like a camp stove)

That's it! With these simple tools, you can create a basic still.

Putting It Together:

1. Fill the pot with the water you want to purify.
2. Place the collection vessel in the center of the pot.
3. Use the upside-down lid or plastic wrap to cover the pot. This will collect the condensed water vapor.
4. Apply heat to start the evaporation and condensation process.

While it may look simple, this setup is a testament to the power of transforming something dirty into something clean.

Using Your Still: Patience is Key

Distillation takes time. The water slowly changes from liquid to vapor and back again. Be patient and keep an eye on your still. Adjust the heat as needed to keep the evaporation happening at a steady rate.

Once you have your distilled water, it's important to store it properly to keep it clean. Use sterilized and sealed glass containers. Distillation is like a purification ritual, transforming water from something unsafe to something drinkable.

Benefits and Limitations

Distillation is a great way to purify almost any water source, making it a very versatile tool. However, there are a few things to keep in mind:

Time and Energy: Distillation is a slow process and requires a constant heat source, which can be a challenge if resources are limited.

Mineral Removal: Distillation removes not only contaminants but also minerals that are good for your body. While the purified water is safe to drink, you may want to consider adding a pinch of salt or some mineral drops to improve the taste and nutrient content for long-term consumption.

Here's a checklist to help you build and use your water still:

- Gather Materials: Pot, collection vessel, lid/plastic wrap, heat source
- Assemble: Put the pot on the heat source, fill it with water, and place the collection vessel in the center.

- Start Distillation: Cover the pot with the lid or plastic wrap (upside down) to collect condensation.
- Monitor: Adjust heat and watch for the water vapor to turn back into liquid.
- Store Distilled Water: Transfer the collected water to sterilized and sealed glass containers.
- Consider Mineralization: You may want to add some minerals back into the water for better taste and nutrition.

With these steps, you can transform the act of water purification into a process of self-reliance. Distillation not only teaches you a valuable skill but also embodies the spirit of resourcefulness and independence that's essential for survival.

7.2 URBAN RAINWATER HARVESTING: A GUIDE FOR CITY PREPPERS

In cramped city environments, capturing rainwater becomes a solution for water scarcity and a clever way to reduce reliance on municipal systems. This section dives into rainwater harvesting for urban preppers, navigating limited space and maximizing yield from rooftops. Here's what you need to know about selecting components, assembling a compact system, ensuring water purity, and adhering to regulations.

Urban Considerations for Rainwater Harvesting

Space constraints in cities demand creative solutions. Extensive collection areas are impractical. Instead, urban rainwater harvesting systems leverage the verticality of buildings.

Gutters strategically positioned on rooftops become the first step, channeling rainwater into storage barrels. These barrels need to be a balance of slim and spacious, unobtrusive yet accessible. Filters then remove debris and leaves, keeping the collected water clean.

Assembling Your Urban Water System

1. Strategic Gutter Placement: Proper gutter placement is crucial to collect maximum rainwater. Ensure gutters are installed correctly to avoid roof damage and guarantee smooth water flow.
2. Connecting the System: Connect gutters to barrels using downspouts, which channel water from the gutters to the storage containers.
3. Filtering for Cleanliness: Install filters at the connection point between the downspout and the barrel to remove debris from the runoff.
4. Double Filtration: For enhanced protection, consider a first flush diverter. This device discards the initial rainwater flow, which typically contains the most pollutants from the roof. Only the cleaner rainwater afterward is diverted into your barrels.

Making Rainwater Usable

Collecting rainwater is just one step. Safe use, especially for drinking, requires additional purification:

- UV Treatment and Sedimentation: After collection, UV light and sedimentation can eliminate harmful bacteria and allow remaining particles to settle. This two-pronged approach transforms rainwater from a basic collection project into a safe and valuable water source.

Responsible Urban Prepping

- Local Regulations: Before you start collecting rainwater, check your local bylaws and codes. These may outline legal requirements for rainwater harvesting in your city. Following the rules ensures your system is safe and compliant with community standards.
- Safety First: Ensure your roof can support the extra weight of gutters and full barrels. Your design should also prevent water contamination. Prioritizing safety protects you and promotes responsible resource management in your city.

Rainwater: A Step Towards Urban Resilience

With these techniques, you can create a rainwater harvesting system in your urban environment. This system is more than just collecting water; it's about becoming self-reliant and lessening dependence on the municipal water supply.

It's also a way to reduce your household's environmental impact. Even in a concrete jungle, you can find ways to coexist with nature and harness its resources responsibly.

7.3: TURNING TRASH INTO TREASURE: GREYWATER SYSTEMS FOR SURVIVAL

Greywater, the wastewater from your sink, shower, and washing machine, might seem like something to discard. But in a survival situation, it becomes a hidden gem - a valuable resource waiting to be tapped. Unlike toilet water (blackwater), greywater isn't as contaminated, but it's not suitable for drinking either. However, with a little ingenuity, you can transform this "waste" into a source of water for non-potable needs.

The Power of Reuse: Every Drop Counts

Greywater allows you to stretch your existing water supply further. The water you use for washing your body or dishes can be redirected to water your garden or flush toilets. This reduces waste and maximizes the use of every precious drop – a key principle for sustainable survival.

Important Considerations: Not a One-Size-Fits-All Solution

While greywater offers a valuable resource, it's important to understand its limitations:

- Restricted Uses: Greywater shouldn't be used for everything. Avoid using it on edible plants or in ways that could lead to direct human contact.
- Mind the Contamination: Greywater can still contain some contaminants like soap residue. Be cautious and use appropriate methods to minimize risks.

Simple Greywater Systems: DIY Water Conservation

The beauty of greywater systems lies in their simplicity. Forget fancy technology – here are a couple of examples you can set up yourself:

- Washing Machine to Reed Bed: Divert water from your washing machine to a constructed wetland filled with reeds. These plants act as natural filters, purifying the water for irrigation.

- Shower to Mulch Basin: Direct shower water to a basin filled with mulch. The mulch helps filter the water while providing moisture for your plants. The soil beneath the mulch acts as an additional natural filter, further reducing any remaining contaminants.

Safety First: Essential Precautions

When using greywater, safety is paramount. Here are some key points to remember:

- Prevent Backflow: Ensure your greywater system doesn't accidentally contaminate your clean drinking water supply.
- Smart Irrigation: Don't use greywater directly on edible plants. Opt for subsurface irrigation to minimize human contact with the water.
- Treat the Water: Even simple greywater systems should address bacterial growth. Don't let the stored water become a breeding ground for harmful bacteria.

The Rules of the Game: Navigating Local Regulations

Laws and regulations regarding greywater use can vary depending on your location. Some areas encourage greywater use with specific guidelines, while others have restrictions in place. Always check your local laws before setting up a greywater system.

Beyond the Law: Environmental Benefits

Greywater reuse isn't just about saving water for yourself. It also benefits the environment in several ways:

- Reduced Wastewater: Less greywater entering sewage systems means less stress on local water treatment plants and natural water bodies.
- Smaller Footprint: By reusing water, you reduce your overall water consumption, minimizing the environmental impact of your household.

A Sustainable Approach to Survival: More Than Just Staying Alive

Greywater systems represent a testament to the idea that water is a precious resource, not to be wasted. They embody a broader philosophy of survival that goes beyond simply staying alive.

By using greywater wisely, we not only gain an additional water source but also demonstrate our commitment to preserving the delicate balance of our environment. In their simplicity and effectiveness, greywater systems become symbols of a sustainable future, one where we learn to use and reuse water responsibly.

7.4: BUILDING YOUR OWN SOLAR WATER PURIFIER: A DIY GUIDE

The sun – a powerful and free source of energy - can be harnessed to create clean drinking water. Solar distillation is a brilliant and sustainable way to achieve this. It uses sunlight to transform dirty water into its purest form, with minimal equipment.

Simple Yet Effective: The Power of the Sun

Solar distillation is all about using the sun's heat and light for purification. Contaminated water is placed in a special container and heated by the sun. The water evaporates, leaving contaminants behind. This vapor then condenses on a cool surface and drips down as clean, purified water.

Building Your Solar Still: No Fancy Tools Required

The beauty of a solar still is that you can build it yourself with basic materials:

Black Shallow Pan: This acts as the evaporation chamber. Black absorbs heat best, so choose a dark pan to maximize the sun's effect.

Transparent Cover: A clear sheet of glass or plastic will go over the pan. Make sure it creates an airtight seal around the edges.

Location is Important:

Where you place your solar still is important for optimal performance:

- Sun Exposure: Find a location that receives direct sunlight for most of the day. In the Northern Hemisphere, this usually means facing south.
- Sun's Angle: For best results, tilt the clear cover at the same angle as your latitude. This ensures the sunlight hits the pan directly.

Boosting Efficiency:

Here are a few tricks to get even more clean water from your solar still:

Reflectors: Line the outside of the pan with aluminum foil or mirrors. These reflect sunlight onto the pan, increasing the heat.

Regular Cleaning: Clean the pan regularly to remove any residue that might block sunlight. Also, keep the clear cover clean to ensure proper condensation.

Maintaining the Seal: Check the seal around the edges of the pan and cover for leaks. A good seal is essential to prevent vapor from escaping.

Troubleshooting Your Solar Still:

If you're not getting as much clean water as expected, here are some things to consider:

- Less Sunlight: Seasonal changes or cloudy weather can reduce efficiency. Remember, your solar still relies on sunshine.
- Leaks: Check for leaks in the pan or cover. A leaky system loses valuable vapor.

A Sustainable Solution:

A solar still is a perfect example of human ingenuity working with nature. By carefully designing, placing, and maintaining your still, you can tap into the sun's power to create clean water, all while minimizing your environmental impact.

The Bigger Picture: Sustainability and Resourcefulness

Solar distillation is a prime example of using natural processes for water purification. Building and using solar still is a practical application of renewable energy for a vital resource - clean drinking water.

It's not just about immediate needs; it's about promoting environmental responsibility and self-sufficiency in a world with limited resources. The knowledge gained from solar distillation opens doors to other ways of harnessing nature's power as we strive for independence and sustainability.

CHAPTER 8

NAVIGATING THE LEGAL CURRENTS: A PREPPER'S GUIDE TO WATER FOR SURVIVAL

W ater is life. It flows freely in nature, but when we try to capture it for ourselves, things get complicated by laws. This chapter dives into water rights, which are the rules that govern how people can use water. Understanding these rules is important for any prepper who wants to be self-sufficient and have a reliable water source.

8.1 WATER RIGHTS 101

There are two main types of water rights:

- Riparian Rights: Imagine a river running through your property. In areas with riparian rights, owning land next to a water source (like a river or lake) automatically gives you the right to use some of that water. It's like having a built-in water access perk for your land.
- Prior Appropriation: Here, the first person to claim water rights wins. It's like "finders keepers" for water. Whoever

starts using the water first gets the legal right to keep using it, regardless of where their land is.

For a prepper, this makes a big difference. In a riparian state, living near a river or lake is a good way to secure water access. But in a prior appropriation state, the race is on to claim water rights first, often through a complex government process.

Research is Key

Knowing the water laws in your area is crucial. Every state and city has its own rules about how much water you can use, how you can collect it (like rainwater harvesting), and where you can store it. Think of it like a user guide for water.

There are several ways to find this information:

- Government Websites: Many state and local governments have websites that explain water rights and regulations.
- Legal Databases: Online legal resources can provide detailed information on water laws.
- Water Rights Attorneys: If things get confusing, a lawyer specializing in water rights can give you specific advice for your situation.

By doing your research, you'll be sure your water prepping plans are legal and don't break any rules.

Talking it Out: Negotiation can be Your Friend

In some places, water rights are very strict. But that doesn't mean there's no room for negotiation.

Here are some ways to talk your way to a water solution:

- Permits: You might need a permit to drill a well or build a rainwater collection system. Negotiation can help you get the permit approved.
- Sharing is Caring: If you can't get your own water rights, consider joining forces with others in your community. Sharing water rights can be a win-win situation for everyone.

Real-Life Water Right Stories

Water rights can be a source of conflict, but they can also bring people together. Here are some examples:

Colorado Cooperation: In a state with prior appropriation, a small community faced limited water access. They banded together to form a water cooperative. By pooling resources, they were able to navigate the state's water allocation system and secure rights to water from a nearby river. Teamwork made the water dream work!

Virginia Victory: A homesteader wanted to collect rainwater, but local laws didn't allow it. They researched the law, talked to the zoning board, and explained how their system would be sustainable and wouldn't cause any problems. The board listened and eventually approved the rainwater system. This shows how understanding the law and talking it out can lead to success.

Water Rights Checklist for Preppers

Here's a quick guide to help you navigate the water rights maze:

1. Identify Your Source: Is your water coming from a river, underground well, or collected rainwater? Knowing this helps determine what rules apply.
2. Know Your Rights System: Find out if your area uses riparian rights or prior appropriation.
3. Read the Fine Print: Check state and local laws for specific rules on water use, collection, and storage.
4. Talk to the Experts: Consult with water rights attorneys or your state water board if the laws are confusing.
5. Paperwork Power: Gather documents for permits or water rights transfers. Be prepared to negotiate terms that follow the law and are fair to your community.
6. Stay Updated: Water laws can change, so keep yourself informed about any updates.

By following these steps, you can ensure your water prepping efforts are legal and sustainable.

8.2: ETHICAL WATER COLLECTION FOR LONG-TERM SURVIVAL

Prepping for survival often centers on securing water, but how we collect it matters just as much. It's not just about filling your bucket; it's about doing so responsibly. This chapter explores ethical water collection, where survival needs are met while also protecting the environment for the future.

The Responsibility of Water: A Balancing Act

Taking water isn't just about you or your family. It's about respecting the natural world and recognizing water as a shared resource. We need to find a balance between getting life-giving water and keeping nature healthy. This means being thoughtful about how we collect water.

Ethical Water Collection Principles:

1. Minimize Your Impact: Don't take more water than you need, and choose methods that don't harm the environment where you get it. Think of water as part of a bigger system, not something you can just take without consequences.
2. Respect the Cycle: Some methods, like rainwater harvesting, collect water without disrupting the natural flow. This is ideal because it keeps the water cycle in balance.
3. Consider Consequences: Taking too much water from streams, rivers, or underground sources can hurt the plants and animals that rely on it. It can also affect the land itself, causing problems like sinkholes.

Making Wise Choices:

- Rainwater Harvesting: This captures rainwater for later use. It's a great way to get water without affecting the environment.
- Greywater Systems: This reuses water from showers, sinks, and washing machines for things like watering plants. It saves clean water for drinking and cooking.

Community Matters:

In areas with limited water, everyone's actions affect everyone else. Here's how to be a good neighbor:

Work Together: Share knowledge and resources with your community. Maybe you can build a community rainwater system or learn water purification techniques together.

Think Collectively: Plan ways to collect and store water that benefit everyone, not just yourself. Water access should be a shared right, not a competition.

Sustainable Practices: Small Steps, Big Impact

Here are some ways to make your water collection more eco-friendly:

- Use Recycled Materials: Build your rainwater system with things like old buckets or barrels. This reduces waste.
- Natural Filtration: Bio-sand filters use sand and natural processes to clean water. This avoids harsh chemicals.
- Conserve While You Collect: Fix leaky faucets, use water-saving devices in your home, and take shorter showers. Every drop counts!

Adapting for the Future:

Ethical water collection is an ongoing process. Be willing to learn new things and adjust your methods as needed. Remember, water is precious. By respecting its value and its connection to everything around us, we can ensure survival for ourselves and future generations, all while living in harmony with nature.

8.3: THE RIPPLE EFFECT: HOW WE COLLECT WATER IMPACTS OUR ENVIRONMENT

Water is the lifeblood of our local ecosystems. From towering trees to tiny insects, everything relies on this precious resource. But when we take water for ourselves, it can have a ripple effect, changing the complex web of life in these ecosystems.

Creating a Balance Between Water and Ecosystems

Healthy ecosystems function like well-oiled machines. Streams and rivers flow freely, nourishing plants and animals on land and in the water. Wetlands act as natural filters, cleaning water and providing homes for a variety of birds, fish, and insects. Even the soil needs water to break down nutrients that plants use to grow.

When we take water out of this system, like pumping from wells or diverting streams, it disrupts this balance. We may take out water faster than nature can replenish it, harming the relationships between different plants and animals.

Minimizing Our Impact

There are ways to collect water with less impact. Here are a few ideas:

- Use Less Water: Low-flow faucets and showerheads, along with rainwater harvesting systems, all help reduce the amount of water we take from nature.
- Water-Wise Landscaping: Xeriscaping uses plants that need little to no extra watering, reducing the strain on local water sources.

These strategies give ecosystems a chance to recover and keep functioning properly. This, in turn, helps plants and animals thrive, keeping our environment healthy and resilient.

Water Collection Can Help Too!

Believe it or not, collecting water can actually benefit ecosystems in some cases. Here's how:

Rain Gardens: These gardens are built to capture rainwater runoff and filter pollutants. They provide a mini-habitat for insects, birds, and small animals, all while reducing the burden on storm drains.

Water Features: Artificial ponds or water features filled with collected rainwater can be lifesavers for wildlife during dry periods. They offer a much-needed drink for animals and help maintain healthy populations.

Looking around the world, we can see both positive and negative impacts of water collection. Here are two examples:

- Ogallala Aquifer: In the dry parts of the United States, over-pumping groundwater for farming has caused the Ogallala Aquifer to shrink. This not only hurts farmers who rely on this water but also dries up nearby wetlands and streams, harming the entire ecosystem.
- Johads of Rajasthan: In India, they've revived an ancient practice of building small dams called johads. These dams capture rainwater and replenish groundwater. This simple technique has turned deserts green again and brought back wildlife that had disappeared.

These stories show how important it is to be thoughtful about how we collect water. Our actions can have a big impact on the environment.

The Path Forward: A Symbiotic Relationship

By understanding how water collection affects ecosystems, we can learn to live in harmony with nature. Here's what we can do:

- Be Water Wise: Use water conservation methods and collect rainwater whenever possible.
- Protect Habitats: Support efforts to restore natural habitats and promote biodiversity.
- Respect the Web of Life: Remember, everything is connected. By taking care of the environment, we take care of ourselves.

With these steps, we can ensure that our water needs are met in a way that safeguards the ecosystems that sustain us all. It's about finding a balance, where our survival enriches the planet, not diminishes it.

8.4: RAINWATER HARVESTING LAWS: A STATE-BY-STATE MAZE

Rainwater harvesting laws in the US are a jumbled mess, reflecting how differently states view this age-old yet innovative way to collect water. Depending on where you live, catching rainwater can be anything from a breeze to a bureaucratic nightmare. This is why preppers need to be both careful and curious when navigating these legal twists and turns.

A Changing Landscape: Colorado as an Example

Take Colorado for instance. Rainwater harvesting used to be tightly controlled under a law called "prior appropriation." But recently, things have changed.

Now, homeowners can collect limited amounts of rainwater, as long as it comes from their roofs and is stored in two barrels that hold a total of 110 gallons or less. This shift shows how some states are starting to see rainwater harvesting as a good way to manage water, especially in areas facing water shortages.

States That Embrace Rainwater Harvesting

On the other hand, some states like Ohio and Illinois make rainwater harvesting super easy. They have few restrictions and even encourage people to collect rainwater by offering things like tax breaks and educational programs. In these states, the law reflects a focus on conservation.

They see rainwater harvesting as a way to both ease the burden on city water systems and empower people to be more self-sufficient.

Hurdles to Jump: Permits and Regulations

In some places, you might need a permit to install a rainwater harvesting system. Often, getting this permit depends on following specific rules to prevent contamination and ensure the water is safe to use.

These rules can cover everything from the materials used to capture the rainwater to how it's filtered and stored. The goal is to minimize health risks while maximizing the benefits of using rainwater.

The Power of Advocacy

Don't underestimate the power of people working together to change restrictive laws. Groups of citizens have used data on the effectiveness and environmental benefits of rainwater harvesting to convince lawmakers to be more supportive.

Successful legal changes often come from people working together to show how rainwater harvesting can conserve water, manage stormwater runoff, and provide a buffer against droughts. There are even stories of communities banding together to push for policies that reflect a modern understanding of how valuable and vulnerable water is.

Finding Help: Resources for Preppers

If you want to collect rainwater legally, there are plenty of resources available to help you. State environmental agencies often have websites with guidelines and FAQs that explain the laws in your area and offer advice on designing and maintaining your system.

Non-profit organizations focused on water conservation and sustainable living can also be a big help. They might offer workshops, toolkits, and tips on how to deal with legal roadblocks and advocate for change.

Legal databases and environmental law centers can provide detailed analyses of water rights and regulations, helping you understand your legal rights regarding rainwater use.

A Legal Landscape in Motion

This chapter is like a roadmap of the legal landscape for rainwater harvesting, with areas of restriction and encouragement. It shows how water laws are constantly changing based on environmental policies, new science, and what citizens are pushing for. With this knowledge, preppers can collect rainwater legally while also promoting sustainability and self-reliance.

Beyond the Law: The Importance of Stewardship

As we move on from the specifics of rainwater harvesting laws, it's important to remember the bigger picture. We've seen the importance of understanding the law, the power of advocacy in shaping environmental policies, and the ongoing need to balance human needs with taking care of the environment.

These are important themes that will guide us as we explore more water survival strategies. We want to find ways to be resilient, follow the law, and respect the natural world all at the same time. With this in mind, let's move forward and explore the connection between survival, sustainability, and the law.

CHAPTER 9

NAVIGATING THE CHEMICAL MAELSTROM: A PREPPER'S GUIDE

Factories and other industries keep our world running, but they also create hidden hazards. Chemical spills can pollute our water, making it unsafe to drink. This chapter is your guide to surviving such an event. As a prepper, you shouldn't be afraid, but prepared. With knowledge and planning, you can face a chemical spill with courage and confidence.

9.1 WATER WOES: FINDING SAFE WATER AFTER A SPILL

Clean water is vital! Here's how to find it after a spill.

After a chemical spill, forget trusting the once-refreshing stream or well. Those might now be harboring invisible dangers. To find safe water, you need to understand the spill itself. Chemicals, by nature, tend to flow downhill or seep into the ground. Luckily, official reports from environmental agencies can be your lifesaver.

They'll track the affected areas, helping you locate sources likely spared from contamination.

Here are some good options for safe water:

- Rainwater: If it hasn't touched anything contaminated, rainwater collected in clean containers can be a good source of drinking water.
- Upstream sources: Rivers and lakes located higher up, away from the spill zone, might still have clean water.

Protecting Your Prepper Prepares: Safeguarding Your Water Supply

Be prepared, not scared! Here's how to protect your stored water.

If you have a prepper-mindset, you likely store water. That's smart! But to keep it safe before a spill happens, take some steps. Seal your containers tightly with covers made from materials that block chemicals. Regularly checking these seals and covers ensures they're still working.

For underground wells or tanks, add a secure top to keep out contamination. These simple measures can make a big difference.

When Disaster Strikes: Cleaning Up Contaminated Water

Even with precautions, your water might get dirty. Here's what to do if your stored water gets exposed to chemicals:

Activated carbon filters: These filters act like magnets for certain chemicals, removing harmful molecules from your water.

Reverse osmosis systems: These are more complex but even more effective. They use a special membrane to remove a wider range of contaminants.

Evacuation Essentials: Taking Water With You When You Have to Leave

Sometimes, the best course of action is to get away. If the contamination is too severe, evacuation might be necessary. But don't just run off! Plan your escape route ahead of time, looking for places along the way that have clean water sources. Here are some things to bring:

Portable water filters: These small filters can turn dirty water into safe drinking water.

Water purification tablets: These tablets kill bacteria and other germs in water.

By planning ahead, you'll be prepared to find safe water even if you have to leave your home.

Chemical Spill Survival Checklist:

- Stay informed: Listen to reports about the spill to know affected areas.
- Protect your water: Check and seal your stored water regularly.
- Clean contaminated water: Use filters or reverse osmosis systems if needed.
- Plan your escape: Find evacuation routes with clean water access.
- Be prepared on the go: Pack portable filters or purification tablets.

This guide equips you to navigate the dangers of chemical spills. It shows you what to do and how to think, so you can turn a scary situation into a story of survival.

9.2 NUCLEAR FALLOUT AND SAFE DRINKING WATER: A PREPPER'S GUIDE

Nuclear fallout: When water becomes a threat

A nuclear explosion or reactor accident changes everything, especially how you find safe water. Radioactive dust from the blast, called fallout, settles on everything, including lakes, rivers, and even underground water. This invisible danger makes it crucial to understand how radiation works and how long it lasts.

Fallout and Water Contamination: Understanding the Enemy

Fallout sticks to dust and debris, contaminating water with radioactive particles like iodine-131, cesium-137, and strontium-90. Each one acts differently and stays around for varying times, from days to decades. This is why knowing how long radiation lasts in water is key to finding safe sources.

Finding Safe Water After a Nuclear Fallout

Look deep: Deep wells and aquifers are often shielded from immediate fallout because of their depth. This natural protection makes them good sources of safe water, at least for a while. However, groundwater can move and become contaminated over time.

Be prepared: Sealed, pre-stored water is your safest bet. By planning ahead and storing water, you'll have a clean supply even after a nuclear event.

Decontaminating Water: Making Dirty Water Safe to Drink

Distillation is a great way to remove radioactive particles from water. It works because radioactive particles have higher boiling points than water. When you heat contaminated water, the clean water vaporizes first, leaving the radioactive material behind. However, be careful during distillation to avoid exposure to radioactive steam.

Storing Clean Water: Keeping it Safe

Once you have distilled water, you need to store it carefully to prevent recontamination from radioactive dust still in the air.

Radiation-proof containers: Use containers made of lead or concrete to block radiation. These containers should also have airtight seals to keep out contaminated air.

Regular testing: Regularly check your stored water with a Geiger counter or send samples to labs. This ensures radiation hasn't gotten through your defenses.

Replenishing supplies: Only add water confirmed safe from ongoing monitoring to your reserves. This ensures a clean, sustainable supply.

Knowledge is Power: Facing the Challenge with Confidence

In a nuclear fallout situation, knowledge is your best weapon against fear. Understanding how radiation works and how to find safe water empowers you to survive. The act of finding clean water, even amidst invisible threats, becomes a symbol of resilience and the human spirit's ability to adapt.

With each safe sip of water, you take control of your situation and chart a course towards survival.

9.3 FLOODS AND SAFE WATER: A PREPPER'S GUIDE TO PREPARATION AND RECOVERY

Floods are powerful forces of nature, reshaping landscapes and lives. But even against such odds, with knowledge and planning, you can protect yourself, especially when it comes to safe drinking water. This guide equips you with the strategies to prepare for and respond to flood-related water contamination.

Before the Flood: Securing Your Water Supply

Preparation is key. Understanding weather patterns is important, but safeguarding your water supply is crucial. Stockpile emergency water in advance. Invest in sturdy, waterproof containers that can withstand floodwaters. These containers will shield your water from contaminants like sewage, chemicals, and debris unleashed by the flood.

Innovation is also your friend. Consider using floatable water storage solutions. These buoyant containers, clearly marked and securely anchored, will bob above the flood, ensuring you have a reserve of clean water once the waters recede.

After the Flood: Assessing Water Safety

As floodwaters recede, a new challenge arises: ensuring your water is safe to drink. The once-familiar landscape might be altered, and invisible threats may now lurk in your water sources. Be vigilant for signs of contamination. Look for unusual smells or discolored water, which could indicate the presence of pollutants from flood debris and sediment.

Don't rely solely on sight and smell. Water testing kits are essential tools. They help you identify wells and springs that haven't been contaminated and differentiate them from compromised water sources.

Making Flood Water Safe to Drink

Floods often leave water contaminated with biological and chemical hazards. To address this, a two-pronged approach of filtration and purification is necessary. This tackles both visible and invisible threats.

First, use sediment filters. These filters have fine mesh pores that trap particles that make the water cloudy. Following this initial step, consider more advanced filtration units. These units employ activated carbon and reverse osmosis membranes to remove toxins more precisely.

However, the battle isn't over yet. Biological threats can still be present. Use UV light purifiers or chemical disinfectants to eliminate them, ensuring your water is truly safe to drink. This multistage approach ensures thorough water decontamination.

The Road to Recovery: Restoring Water Systems

The journey back to normalcy after a flood involves restoring and rebuilding water systems. Personal wells and community systems alike require careful rehabilitation. This begins with flushing pipes and tanks to remove any flood residues. Next comes sterilization using bleach or iodine solutions to eliminate remaining pathogens. For wells and boreholes, professional evaluation is crucial to address any damage or contamination.

Restoring water systems is a collaborative effort. As individuals and communities work together to rebuild these vital systems, it becomes a testament to their resilience. It reaffirms the bond forged in the face of adversity as people work together to ensure clean water for all.

Using these steps, you can navigate the challenges of flood-related water contamination and ensure your family's safety and well-being.

9.4 DROUGHT SURVIVAL: A PREPPER'S GUIDE TO WATER CONSERVATION AND RESOURCEFULNESS

In drought-stricken regions, the once-reliable rain becomes a distant memory. The land parches, crops wither, and water transforms from an abundant resource to a precious commodity. For those who call these arid landscapes home, conservation becomes a way of life, demanding wisdom and resourcefulness to navigate the challenges of a dry climate.

Water Conservation: Every Drop Counts

The key to drought survival lies in a two-pronged approach: reducing water consumption and maximizing its efficiency. Simple changes in daily habits can make a big difference. Replacing conventional showerheads and faucets with low-flow alternatives reduces water use without sacrificing comfort. Though an initial investment, water-efficient appliances like washing machines and dishwashers pay off in the long run by conserving this vital resource.

Beyond the bathroom and kitchen, water conservation extends to your garden. Applying a layer of mulch around plants helps retain moisture in the soil, reducing the need for frequent watering. Drip irrigation systems deliver water directly to plant roots, minimizing waste from evaporation. These techniques not only conserve water but also promote healthier plant growth.

Thinking Outside the Tap: Exploring Alternative Water Sources

As traditional sources like wells and springs dwindle, the search for alternatives becomes crucial. Innovative solutions like atmospheric water generation offer a lifeline in dry climates. These devices capture humidity from the air and condense it into clean drinking water, a modern marvel that defies the limitations of arid environments.

Similarly, greywater, the wastewater from showers, sinks, and washing machines (excluding water contaminated with toilet waste or cleaning chemicals), can be repurposed for irrigation and landscaping. This practice, often referred to as greywater harvesting, transforms what would otherwise be considered waste into a valuable resource for your garden.

By utilizing greywater responsibly, you not only conserve freshwater but also promote a more sustainable lifestyle.

Rationing with Reason: Prioritizing Essential Needs

In the heart of a drought, rationing water becomes an essential practice. It's a delicate balancing act that prioritizes essential needs while minimizing waste. Health and well-being take center stage, ensuring there's enough water for drinking, cooking, and basic hygiene.

Developing a water rationing plan allows you to allocate specific amounts for daily tasks. Creating a schedule that staggers water-intensive activities, like laundry, helps you maintain discipline and ensure every drop is used with purpose. This mindful approach to consumption fosters a culture of water conservation within your household, teaching everyone the value of this precious resource.

Building a Sustainable Future: Living in Harmony with the Arid Environment

Surviving a drought demands not just temporary solutions, but a long-term vision for sustainable living in an arid climate. Xeriscaping, the practice of landscaping with drought-tolerant plants native to your region, is a cornerstone of this vision. These plants are adapted to survive with minimal irrigation, reducing your reliance on scarce water resources.

Rainwater harvesting, though challenged by the limited rainfall, offers another sustainable solution. By capturing even a small amount of rainwater in barrels or cisterns, you can store it for later use in your garden or for non-drinking purposes around the house.

The strategies outlined here serve as a roadmap to navigating the challenges of drought. By implementing water conservation techniques, exploring alternative water sources, rationing with reason, and adopting sustainable practices, you can build resilience and thrive in even the driest conditions.

CHAPTER 10
NEW IDEAS FOR WATER RESILIENCE

As the environment changes rapidly, finding clean and safe water is more important than ever. New technologies in water filtration and purification are emerging, pushing the boundaries of what's possible. This chapter looks at these cutting-edge technologies that could change how we access and use water, ensuring we can meet future challenges head-on.

10.1 THE FUTURE OF WATER FILTRATION: NEW TECHNOLOGIES

Advanced Filtration Materials

Graphene is an amazing material that's just one atom thick but stronger than steel. It has a honeycomb structure that can trap tiny contaminants like bacteria and heavy metals. Imagine a straw coated with graphene that can turn dirty, bacteria-filled water into clean drinking water instantly. This technology is still being developed but could revolutionize how we access clean water, especially in places without traditional water filtration systems.

Intelligent Filtration Systems

Combining sensors and the Internet of Things (IoT) with filtration systems brings a new level of efficiency. These smart systems can monitor water quality in real-time and adjust filtration settings automatically to remove detected contaminants. For example, in a city, such a system could detect and eliminate a sudden contamination in the water supply immediately, protecting millions of people with minimal human effort.

Energy-Efficient Purification

Sustainable water purification is essential, and solar-powered and bio-inspired systems are leading the way. These systems mimic natural processes like plant transpiration and use solar energy to purify water. Picture a remote desert community using solar power to clean water or a home system that mimics plant roots to filter rainwater collected from the roof. These methods are not only effective but also cheaper and more energy-efficient than traditional methods.

Making Technology Accessible and Scalable

For these new technologies to make a real difference, they must be accessible and scalable. This means making them affordable and easy to use and maintain. For instance, portable graphene filters could be distributed in disaster areas, providing immediate clean water and saving lives. Smart filtration systems could be used in homes and large cities, each customized to meet specific needs.

Practical Steps for Adopting New Water Filtration Technologies

1. Research New Materials: Keep up with advancements like graphene and their potential uses in water filtration.
2. Evaluate Smart Filtration: Look at the benefits of IoT-enabled systems for your home or community and see if they work with your current setup.
3. Explore Solar-Powered Solutions: Check out solar-powered purification systems for their efficiency, cost, and suitability for your area.
4. Consider Bio-Inspired Systems: Investigate systems that mimic natural water purification processes, which can be cheaper and use less energy.
5. Plan for Accessibility: Choose technologies that are easy to use, affordable, and widely available.
6. Focus on Scalability: Ensure the solutions you pick can be scaled up or down to meet the needs of different environments, from rural homes to large cities.

This checklist is a guide for individuals and communities ready to embrace new water filtration technologies. As we move forward, these innovations offer hope for a future where everyone has access to clean water.

10.2 PORTABLE DESALINATION DEVICES: SEAWATER TO DRINKING WATER

As freshwater becomes increasingly scarce, particularly in coastal and arid regions, portable desalination devices offer an innovative solution. These devices, ranging from manual and solar-powered to battery-operated models, transform seawater into drinkable water by removing salt and other minerals.

At their core, these devices use either reverse osmosis or distillation methods. Reverse osmosis devices push seawater through a semipermeable membrane that blocks salt and minerals, while distillation units boil seawater, capturing the steam and condensing it back into liquid form, free of salt and impurities.

The benefits of portable desalination devices in coastal survival scenarios are immense. For individuals or communities living near the ocean, these devices are a lifeline, providing fresh water even when traditional sources are compromised by natural disasters or pollution.

Types of Portable Desalination Devices

Manual Desalination Devices: These devices are powered by hand and are useful in emergency situations where other power sources are unavailable. They are often compact and lightweight, making them ideal for personal use or small groups.

Solar-Powered Desalination Devices: Utilizing the sun's energy, these devices are particularly advantageous in sunny climates. They offer sustainability benefits, as they don't rely on electricity or fossil fuels, making them suitable for remote or disaster-stricken areas where conventional energy sources may be unavailable.

Battery-Operated Desalination Devices: These models provide flexibility and ease of use, especially in locations with limited access to consistent sunlight. However, their effectiveness is contingent on having a reliable method to recharge the batteries, which can be challenging in off-grid scenarios.

Advantages and Applications of Desalinating Seawater

Portable desalination devices are particularly valuable in coastal and survival scenarios:

Emergency Response: In the aftermath of natural disasters, such as hurricanes or tsunamis, traditional freshwater sources can be disrupted. Portable desalination devices can provide immediate access to drinking water, crucial for survival and recovery.

Sustainability: Solar-powered models leverage the sun's abundant energy, ensuring a continuous water supply without needing external power sources. This is particularly beneficial in remote areas where electricity is scarce.

Mobility and Flexibility: Manual and battery-operated devices offer great mobility, allowing individuals to access fresh water on the go. This is especially useful for activities like camping, hiking, and extended expeditions.

Challenges You Might Face!

While portable desalination devices offer numerous benefits, they also come with challenges:

Energy Requirements: Battery-operated models need regular recharging, which can be difficult in off-grid areas. Solar-powered units, while reducing some energy concerns, can be less effective under cloudy conditions or at night.

Brine Disposal: The desalination process produces brine, a highly concentrated salt solution. Improper disposal of brine can harm marine life and ecosystems. Users must plan for responsible brine management, ensuring it is diluted and dispersed in a way that minimizes environmental impact.

Real-World Examples

Case Study 1: Cyclone Aftermath

After a cyclone struck a remote island, cutting off its freshwater supply, solar-powered desalination units were deployed as part of the emergency response. These units provided a crucial source of drinking water, allowing the community to sustain itself until traditional water infrastructure could be restored.

Case Study 2: Kayaking Expedition

A group of kayakers on a long coastal journey used manual desalination pumps to convert seawater into drinking water. This significantly reduced their need to carry heavy water supplies, enabling them to explore further and for longer periods.

Future Potential and Continued Development

The success stories of portable desalination devices highlight their practicality and effectiveness in diverse situations. They not only provide essential water supplies in emergencies but also support longer explorations and habitation in remote coastal areas.

- Innovation and Refinement: Ongoing research and development aim to enhance the efficiency and affordability of these devices. Advances in materials, such as graphene, could lead to even more effective filtration processes.
- Sustainability Efforts: Future designs will likely focus on reducing energy consumption and improving brine management to minimize environmental impact. Solar-powered models, in particular, will continue to be refined to work more effectively under varying weather conditions.

- Wider Accessibility: Efforts to make these devices more affordable and accessible will be crucial. This includes developing models that are easy to use and maintain, ensuring they can be deployed quickly in emergency situations or used by individuals without specialized training.

Portable desalination devices are transforming how we access clean water, particularly in coastal and disaster-prone areas. By providing a reliable source of fresh water, they support both emergency response and everyday survival in challenging environments. As these technologies continue to evolve, they hold the promise of further alleviating freshwater scarcity, offering a beacon of hope for communities worldwide.

10.3 THE ROLE OF AI IN WATER PURIFICATION: CUTTING-EDGE ADVANCES

Technology is crucial in improving water purification, and artificial intelligence (AI) is leading the way. AI is changing how we manage and clean water, making processes more efficient and precise. This chapter explores how AI is transforming water purification, ensuring better water quality and safety.

Monitoring Water Quality with AI

AI is revolutionizing how we monitor water quality. Traditional methods often involve periodic testing and can miss sudden changes or emerging contaminants. In contrast, AI uses data from various sensors placed in water bodies and pipelines to continuously analyze water quality. These sensors collect information on different parameters like temperature, pH, and the presence of specific contaminants.

AI algorithms process this data in real-time, detecting contaminants from the visible to the molecular level. This high level of accuracy surpasses traditional methods and allows for immediate identification of any water quality issues. For instance, if harmful bacteria or chemicals are detected, AI can alert authorities instantly, ensuring quick response times.

Predictive Capabilities

Beyond just detecting problems, AI can predict potential issues before they become serious. By analyzing trends and patterns in the data, AI can forecast contamination events. For example, if there is a pattern of increasing chemical levels in a river, AI can predict when these levels might reach dangerous thresholds.

This predictive ability allows for proactive measures to be taken, such as adjusting treatment processes or issuing warnings to communities.

In areas with significant industrial activity or poor sanitation, the risk of water contamination is high. AI's predictive capabilities are especially valuable in these regions, helping to prevent waterborne diseases and chemical pollution before they pose a threat to public health.

Automated and Adaptive Purification Systems

AI is also being used to develop automated purification systems that adapt to changing water quality conditions. Traditional water treatment systems operate on fixed schedules and processes, which may not always be efficient. In contrast, AI-driven systems adjust filtration and purification methods in real-time based on the detected contaminants.

For example, if an AI system detects an increase in organic pollutants, it can automatically increase the intensity of carbon filtration stages to effectively remove these contaminants. During times when water quality is relatively high, the system can reduce its activity, conserving energy and extending the lifespan of the filtration media.

This dynamic adjustment ensures that water is always treated efficiently and effectively, regardless of fluctuations in quality.

Predictive Maintenance

AI's benefits extend to the maintenance of water purification infrastructure. Traditional maintenance schedules are often based on fixed intervals, which can lead to either premature maintenance or unexpected failures. AI changes this by predicting when maintenance is needed based on data analysis.

By continuously monitoring the performance of purification systems, AI can identify signs of wear and tear or potential failures. It can then schedule maintenance before a breakdown occurs, reducing downtime and ensuring a consistent supply of clean water. This is particularly important in resource-limited areas where water systems are crucial for survival.

Case Studies and Real-World Applications

The impact of AI in water purification is evident in several real-world applications. In urban areas, AI-driven water treatment systems optimize processes, reducing the environmental footprint of cities. For example, AI can help manage the energy use of purification plants, balancing high performance with low environmental impact.

In remote or rural areas, AI-powered purification units can transform contaminated or brackish water into safe drinking water. These units are particularly beneficial in regions with limited access to clean water, improving health and quality of life.

Broader Implications

AI's role in water purification goes beyond immediate benefits. It represents a significant step towards global water safety and accessibility. By improving water quality standards, AI helps reduce the disparity between different regions. Imagine remote villages where AI systems provide reliable, clean water or urban centers where optimized water treatment processes support large populations sustainably.

The democratization of water safety through AI can significantly impact the fight against waterborne diseases and environmental degradation. As AI technology continues to evolve, its potential to address water scarcity and pollution will only grow, offering hope for a future where clean, safe water is a universal right.

The integration of AI in water purification is transforming how we ensure water quality and safety. From real-time monitoring and predictive capabilities to automated purification and maintenance, AI brings unprecedented efficiency and precision.

10.4: CROWDSOURCING SOLUTIONS - COMMUNITY INNOVATION IN WATER SURVIVAL

From Individual Struggle to Collective Brilliance: The Power of Crowdsourcing

The fight for water security is no longer a solitary battle. A powerful shift is happening – a move towards crowdsourcing, where communities are pooling their knowledge and creativity to overcome water challenges. This collaborative approach unlocks a treasure trove of grassroots solutions, creating a rich tapestry of techniques for securing this vital resource.

Through online platforms and local projects, people from all walks of life are combining their experiences and ingenuity to make advancements in water purification, conservation, and sourcing.

Digital Platforms: A Meeting Ground for Minds

Platforms like the "Global Water Network" and "AquaShare" exemplify the digital commons where innovation thrives. These online spaces act as virtual town squares, buzzing with ideas. Here, low-tech filtration solutions designed for remote villages mingle with sophisticated rainwater harvesting systems built for urban resilience.

A retired engineer might share a blueprint for a DIY solar distiller, while a rural farmer contributes their time-tested knowledge of constructing sand and gravel filters.

Every contribution, be it a detailed technical drawing or a simple hack passed down through generations, enriches the collective pool of knowledge. This open exchange fosters a culture of innovation where everyone has a voice and every idea holds value.

From Idea to Impact: Real-World Examples of Crowdsourced Success

The impact of these collaborative efforts extends far beyond the digital realm. Countless communities around the world are experiencing tangible improvements in their water security thanks to crowdsourcing. In one Southeast Asian town plagued by arsenic-contaminated groundwater, a simple yet effective arsenic removal method using readily available local materials was discovered through a crowdsourcing platform.

This solution not only brought immediate relief to the community, but also became a model for other regions facing similar challenges.

Another inspiring example comes from sub-Saharan Africa, where shared designs for rainwater catchment systems have significantly reduced water scarcity. Previously arid areas are now transformed into thriving agricultural land, all thanks to the power of shared knowledge.

Fueling the Fire: Encouraging Participation and Growth

For crowdsourcing initiatives to flourish, active participation is key. Incentives play a crucial role in attracting and motivating contributors. Recognition and awards for innovative solutions celebrate creativity and inspire further problem-solving. Additionally, offering implementation support for the most impactful ideas bridges the gap between concept and reality, turning promising ideas into tangible solutions.

Education is another critical pillar. Both formal and informal educational programs equip participants with the knowledge they need to contribute meaningfully. Workshops, online webinars, and readily available resources create an environment of continuous learning, ensuring everyone has the tools to be a valuable part of the solution.

Community engagement initiatives like local workshops and online forums further strengthen collaboration by providing platforms for interaction and knowledge exchange.

Crowdsourcing is not just a trend; it's a paradigm shift in how we approach water challenges. By harnessing the collective ingenuity of communities worldwide, we can unlock a future where everyone has access to safe, clean water.

CHAPTER 11

NURTURING VITALITY: THE CRUCIAL ROLE OF HYDRATION

Have you ever felt sluggish, foggy-headed, or just plain out of it? Dehydration, simply not drinking enough fluids, could be the culprit. It can sneak up on you without you even realizing it. This chapter will explore why water is so important and how to make sure you're getting enough throughout the day.

11.1 THE EFFECTS OF DEHYDRATION ON THE BODY AND MIND

Water is like the foundation of your body. It's in every single cell, making up a large portion of your weight. Just like a car needs oil to run smoothly, your body needs water for almost everything it does, from flushing out waste products to regulating temperature. When you don't drink enough water, even a little bit, it can throw a wrench into the system.

Imagine your kidneys, which act like a filter, struggling to do their job because of low water levels. Waste products start to build up, and that can make you feel sluggish and unwell. Dehydration can also mess with your body's ability to control its temperature. Normally, you sweat to cool down, but if you're dehydrated, you can't sweat as much. This can lead to overheating and even a serious condition called heatstroke.

Dehydration and Your Brain: When Thinking Gets Cloudy

Your brain is the control center of your body, calling the shots and keeping everything working together. But just like any good manager, your brain needs the right tools to function well. One of the most important tools? Water! When you're dehydrated, your brain doesn't work as well as it should.

It might be harder to concentrate, pay attention, or remember things. You might also feel foggy-headed and just not quite yourself. So, staying hydrated is essential for keeping your brain sharp and helping you think clearly throughout the day.

How to Spot Dehydration's Signals

Most of the time, your body is pretty good at letting you know when it needs something. When you're hungry, your stomach growls. When you're tired, you yawn. And when you're dehydrated, you feel thirsty! Thirst is your body's way of saying, "Hey, I'm running low on water, I need a refill!" But here's the catch: sometimes thirst doesn't kick in right away. By the time you feel really thirsty, you might already be a little dehydrated.

There are other signs to watch out for too. If your mouth feels dry and sticky, that's a clue you might need some water. Also, pay attention to the color of your pee. Pale yellow is a good sign, but if it's dark yellow, that means you're dehydrated and need to dilute it with some water.

Who Needs to Pay Extra Attention to Hydration?

Dehydration can be especially dangerous for some people. Young children, for example, might not be able to tell you they're thirsty, so it's important for parents and caregivers to watch out for signs like a fever or fussiness. Older adults can also be at risk because their bodies might not be as good at sensing thirst. People with certain health conditions might also be more prone to dehydration. If you're taking care of someone who falls into one of these categories, it's important to be extra vigilant about making sure they drink enough fluids throughout the day.

Your Daily Dose of Hydration: Simple Tips for Staying Hydrated

So, how can you make sure you're getting enough water throughout the day? Here are some easy tips to follow:

- The Morning Check-In: When you wake up in the morning, take a moment to see if you're thirsty. Also, check the color of your pee. Pale yellow is a good sign of hydration, while dark yellow means you might be dehydrated and need to start your day with a big glass of water!
- Sip Throughout the Day: Don't wait until you're parched to drink water! Instead, try to sip on water regularly throughout the day. Think of it like watering a plant - a little bit often is better than a whole lot at once. You can set reminders on your phone or keep a reusable water bottle with you as a reminder to take a sip every now and then.
- Water with Every Meal: Water doesn't have to be the only thing you drink. Many fruits and vegetables are actually

packed with water, like watermelon, cucumber, and strawberries. Including these in your meals and snacks is a great way to boost your hydration levels.

11.2: STAYING HYDRATED IN TOUGH SITUATIONS

Staying hydrated is crucial, especially when you're facing scorching heat or pushing your body hard. Water acts as your lifeline, helping you deal with both the demands of the environment and the strain on your body. But in these tough situations, every sip counts. It becomes a balancing act between staying hydrated and making the most of what little water you might have.

Adapting to the Environment

The key to staying hydrated in harsh environments is understanding your body's signals and the conditions around you.

In hot climates, the sun sucks moisture out of your skin with every breath you take. To fight this invisible enemy, you need to drink more fluids. But it's not just about how much you drink, it's also about when you drink.

Here's a winning strategy for hot weather:

Drink before it gets hot: Get ahead of the heat by downing plenty of water before the temperature soars.

Sip often while resting: Take regular sips throughout the day, especially when resting in the shade.

Replenish at the end of the day: End your day by rehydrating to replace the fluids you lost through sweating.

This way, you're giving your body a steady supply of water throughout the day.

Making the Most of Scarce Water

Now, imagine a situation where water is scarce and every drop is precious. Here, you need to make the most of what you have. You might need to become a water rationing pro:

Small sips: Take small sips instead of gulps to stretch your water supply.

Careful use: Carefully consider how much water you use for things like washing yourself or cooking.

Getting Creative with Water Collection

In these situations, you can even get creative to find more water:

Condensation collection: Collect condensation that forms on objects at night, like wringing out a wet cloth that was left outside.

Morning dew harvest: Try to gather morning dew from plants.

These methods might seem unusual, but they show just how important water is for survival. People have found ways to get water even in the harshest places.

Boosting Hydration with Special Aids

There are also special products that can help you stay hydrated in tough situations. These hydration aids come in powder or tablet form and contain things like sodium, potassium, and sugar. They work by mimicking the natural way your body absorbs fluids, helping you get more out of every sip. This is especially helpful when you're sweating a lot and losing electrolytes (salts and minerals) through exertion.

By using these hydration aids, you're helping your body stay strong and healthy even when it's under stress.

Building Mental Strength for Hydration

However, staying hydrated in difficult environments isn't just about physical strategies. It's also a mental battle. It takes willpower to choose water over ignoring your thirst, especially when you're facing other challenges.

Here are some tricks to build your mental strength for hydration:

Carry a reusable water bottle: This serves as a constant reminder to take sips throughout the day.

Set water alarms: Use your phone's alarm to remind yourself to drink water at regular intervals.

Working Together for Hydration

Another way to stay motivated is to work together with others:

Team effort: Remind each other to drink water, making it a team effort. This makes it easier to stay hydrated, especially when things get tough.

All in all, staying hydrated in high-stress environments isn't just about how much water you drink, it's about adapting to the situation, making the most of what you have, and having the mental strength to keep going. From planning your water intake to using special aids and staying positive, there are many ways to make sure water remains your source of strength as you overcome the challenges of tough environments.

11.3: STAYING SAFE FROM WATERBORNE ILLNESSES

When we're out in the wilderness or even inside our shelters, there's a hidden danger lurking in unsafe water. Waterborne diseases can turn this essential resource into a source of sickness. These illnesses come from tiny things you can't see, like bacteria, viruses, and protozoa, that live in contaminated water.

There are many different waterborne diseases, each with its own set of problems. Giardia, for example, can cause diarrhea and stomach cramps, making you feel weak. Cholera strikes fast and hard, causing severe diarrhea and vomiting that can quickly lead to dangerous dehydration.

Hepatitis A attacks the liver, causing tiredness and a yellowish color to the skin (jaundice). These are just a few examples, and others like typhoid fever and dysentery can be just as serious.

The key to staying safe from waterborne diseases is prevention. This means being very careful about the water you drink and how clean your surroundings are.

Purification: Making Water Safe to Drink

The first step in prevention is purification. This means turning dirty water into clean water that's safe to drink. There are a few different ways to do this:

- Boiling: This is the oldest and simplest method. Boiling water for at least one minute kills most bacteria and viruses.
- Filtration: Water filters can remove tiny particles and organisms from water. There are many different types of

filters available, so choose one that is appropriate for your situation.

- Disinfection: Using special chemicals like chlorine tablets can also kill bacteria and viruses in water. Be sure to follow the instructions carefully when using disinfectants.

Safe Storage: Keeping Water Clean

Once you have purified your water, it's important to store it safely. This means keeping it in clean, sealed containers to prevent contamination from dirt, insects, or anything else.

Hygiene: Keeping Your Environment Clean

Waterborne diseases can also spread through poor hygiene. This means being careful about how you dispose of waste and keeping your surroundings clean. Here are some tips:

- Always wash your hands with soap and clean water after using the toilet.
- Dispose of waste properly, in a designated area away from your water source.
- Avoid letting water become stagnant (stand still) for long periods, as this can create a breeding ground for germs.

Treatment: Getting Better if You Get Sick

Even if you're careful, there's always a chance you could get sick from contaminated water. If this happens, it's important to seek medical attention right away. Here's how to treat some common waterborne illnesses:

Dehydration: For diseases that cause diarrhea and vomiting, like cholera, oral rehydration solutions (ORS) can be a lifesaver. These solutions contain water, salts, and sugar and help to replenish fluids and electrolytes lost through illness. You can buy ORS at most pharmacies or make your own at home.

Bacterial Infections: For certain bacterial infections, antibiotics can be prescribed by a doctor. It's important to take antibiotics exactly as directed and to finish the entire course of medication, even if you start to feel better.

Building a Community Response Plan

The best way to fight waterborne diseases is to work together as a community. Here's what you can do:

- Share Knowledge: Educate everyone in your group about the dangers of waterborne diseases and how to prevent them.
- Prepare for the Worst: Practice what you would do in case of an outbreak. This could involve things like setting up a water purification system or designating areas for waste disposal.
- Assign Roles: Decide who is responsible for different tasks, such as purifying water, maintaining hygiene, and seeking medical attention if needed.

By working together, you can create a strong defense against waterborne diseases and keep your community healthy. Remember, a little preparation and teamwork can go a long way in protecting yourself and others from these preventable illnesses.

11.4: THE SECRET TEAM BEHIND HYDRATION: ESSENTIAL VITAMINS AND MINERALS

Staying hydrated is crucial for good health, but there's a secret team working behind the scenes to make it all happen: electrolytes! These tiny superheroes might sound fancy, but they're really just special minerals that carry electrical charges throughout your body. They play a vital role in how your body uses water, keeping everything running smoothly like a well-oiled machine.

Electrolytes: The Tiny Messengers of Balance

Imagine your body as a giant city. Every cell is a tiny house, and water is like the delivery trucks bringing supplies. Electrolytes act as messengers, zipping around inside and outside these cells, making sure the water gets delivered to the right place and used effectively. They help maintain a delicate balance of fluids inside and outside your cells, which is essential for good health.

The main players in this electrolyte team are sodium, potassium, magnesium, and calcium. Each one has a specific job:

Sodium: Think of sodium as the traffic controller. It helps regulate the amount of water in your body and keeps the pressure balanced between your blood and cells.

Potassium: This mineral works hand-in-hand with sodium, helping to move nutrients into your cells and waste products out. It also plays a role in muscle function and nerve impulses.

Magnesium: Magnesium is like the maintenance crew, keeping your muscles relaxed and your heart rhythm steady. It also helps your body use other electrolytes effectively.

Calcium: This strong mineral is important for building and maintaining healthy bones and teeth. It also plays a role in muscle function and blood clotting.

By working together, these electrolytes ensure that every cell in your body gets the water it needs to function properly.

Finding Electrolytes in Nature's Bounty

The good news is that you can find these vital electrolytes in delicious foods! Think of it as a treasure hunt for your body. Here are some clues to help you find the hidden riches:

Potassium Power: Leafy green vegetables are overflowing with potassium. So next time you're making a salad, pile on the spinach, kale, or Swiss chard. Avocados and bananas are also excellent sources of this champion electrolyte.

Magnesium Munchies: Nuts and seeds are packed with magnesium. So next time you're looking for a crunchy snack, grab a handful of almonds, cashews, pumpkin seeds, or sunflower seeds!

Calcium Comfort: Dairy products like milk, cheese, and yogurt are well-known sources of calcium. But did you know that leafy green vegetables also contain some calcium?

Supplements for When Nature Runs Low

Sometimes, finding enough electrolytes in food alone can be tricky, especially if you're sweating a lot due to exercise or hot weather, or if you're not eating a balanced diet. In these situations, electrolyte supplements can be helpful. They come in various forms, like tablets, powders, or even sports drinks, and can help replenish your body's stores.

However, it's important to remember that electrolyte supplements are not a replacement for a healthy diet. They should only be used when needed, and it's always best to talk to your doctor before taking any supplements.

Warning Signs of Electrolyte Imbalance

If your electrolyte balance gets thrown off, your body will start sending you some warning signs. Here are a few things to watch out for:

Muscle cramps: Ever had a leg cramp that made you wince in pain? That could be a sign of an electrolyte imbalance. When your electrolytes are out of whack, your muscles can't function properly and might cramp up.

Fatigue: Feeling wiped out all the time, even after a good night's sleep? Electrolyte imbalance can contribute to fatigue because your cells aren't getting the water and nutrients they need to function properly.

Dizziness: Feeling lightheaded or dizzy could be another clue that your electrolytes are out of whack. This can happen because electrolyte imbalance can affect your blood pressure.

Heart rhythm problems: In severe cases, electrolyte imbalance can affect your heart rhythm. This is why it's important to seek medical attention right away if you experience any of these symptoms.

Keeping Your Body in Harmony

Remember, water is like a highway for electrolytes. You need both water and electrolytes to stay hydrated properly.

The key takeaway here is that hydration is about more than just drinking water. It's about choosing a diet that provides your body with the nutrients it needs to use water effectively. By making smart food choices and staying hydrated, you can keep your body functioning at its best and ensure that your team of tiny electrolytes has everything they need to keep you healthy and energized!

CHAPTER 12

THE TOUGH STUFF IN DRY PLACES: WATER, WASTE, AND STAYING HEALTHY

Imagine a place where water isn't like a constantly running faucet, but a precious treasure you have to be careful with. Every drop counts! In these dry lands, keeping things clean and healthy takes some planning and a bit of smarts. This chapter is all about making a sanitation plan for these dry places, where saving water is a big win for everyone's health.

12.1 MAKING A SANITATION PLAN WHEN WATER IS SCARCE

Why is a Sanitation Plan Important?

Just like any good plan for surviving something tough, especially when water is scarce, a sanitation plan is super important. Without lots of water to wash things away, getting rid of waste becomes a big problem. If waste isn't handled right, it can spread sickness and make a healthy place become dangerous.

Building Your Sanitation Plan

Here are the main things you need for a sanitation plan that works in a dry place, where water is like gold:

Getting Rid of Waste: This is the most important part of any sanitation plan. We need ways to throw things away without using a lot of water. Here are some ideas:

- Toilets That Don't Use Water: These toilets are called dry toilets, and they don't need any water to flush. Instead, they use things like sawdust or other natural stuff to break down the waste safely.
- Keeping Waste Separate: Keeping waste from people (like toilets) away from where you live and eat is super important to stop germs from spreading. This means having special places for toilets and keeping food areas clean.

Keeping Yourself Clean: Washing with water might not always be an option, but that doesn't mean we can't be clean! Here are some clever tricks:

- Ash for Washing Hands: The leftover ashes from your cooking fire can actually be used to clean your hands. Ash has special cleaning powers in it naturally.
- Sand for Scrubbing Dishes: Sand can be used to scrub pots and pans instead of water. It's like a natural sandpaper that gets rid of dirt and grime.

Keeping Food Safe: Even without a lot of water, keeping food clean and safe to eat is super important. Here's a cool trick:

- Sunshine Power: Sunlight has strong rays that can kill germs. We can use this sunshine power to clean our utensils and cookware by leaving them out in the sun for a while.

Thinking Outside the Box: Super Smart Sanitation Ideas

When things get tough, people get creative! In dry places, regular ways of dealing with waste take a backseat to super smart solutions:

- Dry Toilets: These are a great alternative to toilets that use lots of water. They use air and natural stuff like sawdust to break down waste without needing water.
- Sunshine Power for Cleanliness: This way of cleaning uses the free and powerful sun to kill germs on our utensils and water, turning a problem (no water) into a solution (using sunshine).

Working Together for a Healthy Community

Keeping things clean isn't a one-person job. By working together as a community, we can create a healthier environment for everyone. Here are some ideas:

- Sharing Toilets: Building and using toilets together as a community is a great way to save resources and make sure everyone has access to a place to go.
- Learning About Staying Clean: Sharing knowledge about good ways to stay clean is key. Workshops and things you can read can help everyone in the community learn how to stay healthy in a dry place.

A Map to Cleanliness in Dry Places

Imagine a guidebook that shows you exactly how to make a sanitation plan for your dry community. This guide would have different sections to help you get started:

- Plan Picture: A drawing showing the important parts of your sanitation plan, like areas for throwing away waste, stations for washing hands, and places to prepare food.
- Step-by-Step Guide: This would be a list that tells you exactly how to set up each part of the plan, from building dry toilets to setting up sunshine cleaning stations.
- Stuff You Need: This section would tell you what things and tools you need to put your plan into action, along with ideas on where to find them nearby or online.
- Getting Everyone Involved: This section would give you tips on how to get everyone in your community interested in the sanitation plan. It would have ideas for workshops, making things you can read about how to stay clean, and even forming a sanitation committee.

This guide is like a roadmap. It helps communities and individuals take complicated ideas and turn them into real actions. With this guide, people living in dry environments can build their own healthy and clean communities, even when water is scarce.

12.2 SAFE DISPOSAL OF HUMAN WASTE WITHOUT RUNNING WATER

In dry places where running water isn't available, getting rid of human waste safely becomes a critical challenge. Improper disposal can spread harmful germs (pathogens) and cause serious health problems. This is because these pathogens can easily cont-aminate nearby water sources, turning them from sources of life into sources of disease.

This contamination can lead to a variety of illnesses, ranging from stomachaches and diarrhea to fevers and chills. In severe cases, it can even be deadly.

Therefore, finding a safe way to dispose of human waste in dry places is crucial. It's a balancing act - we need to protect people's health while also keeping the environment clean.

The Dangers of Improper Disposal

Think about it like this: if you throw trash in your own backyard and leave it there, what happens? It starts to smell bad, and insects and other pests are attracted to it. This is similar to what happens when waste isn't disposed of properly in dry places. The germs in the waste spread and make people sick.

Smart Solutions for Safe Disposal

Here are some clever ways to dispose of human waste without using a lot of water:

Composting Toilets: These innovative toilets are like magic! They use natural processes like decomposition and evaporation to break down waste into compost. This compost is a valuable soil amend-ment that helps plants grow. The best part? They don't need running water to function – they rely on good bacteria to do the job. This makes them a sustainable option for communities in dry areas.

Cat-Hole Latrines: These are simpler toilets, often used in tempo-rary camps or remote locations. They're basically holes dug in the ground, but there's a specific way to dig them for safety. The goal is to keep the waste far away from water sources and living areas to prevent the spread of germs.

Planning for Success

Just like building a house, planning how to dispose of waste in a dry place is essential. Here are some key factors to consider:

Location is important: Where you place the waste disposal system matters a lot. It should be on higher ground, away from any water sources like wells or springs, and in an area with good drainage. This ensures that when the waste breaks down, it doesn't contaminate anything important.

Considering the Soil: The type of soil also plays a role. Sandy soil allows for better drainage, which is ideal for waste disposal. Clayey soil, on the other hand, doesn't drain as well and isn't the best choice.

Monitoring and Maintenance

Once you have a waste disposal system in place, it's important to keep an eye on it and make sure it's working properly. This means checking on it regularly to ensure the waste is breaking down as expected.

Different Places, Different Solutions

The best way to dispose of waste in a dry place depends on the specific location. In a crowded city, a large composting facility serving the entire community might be the most practical solution. This allows everyone to benefit from a sustainable waste management system.

In a remote wilderness area, on the other hand, a simple cat-hole latrine system, spaced far apart from each other, might be the best approach. This minimizes the impact on the pristine environment.

From Challenge to Harmony

By using smart methods like composting toilets and cat-hole latrines, we can safely dispose of human waste even without running water. These solutions are not just practical answers to a challenge, but they also show how humans can adapt and live in harmony with the environment, even when resources are scarce.

Remember, proper waste disposal is key to keeping our communities healthy and protecting the environment for future generations.

12.3 STAYING CLEAN WHEN WATER RUNS DRY: A BALANCING ACT OF NECESSITY AND INNOVATION

In places where water is a precious treasure, not a constant flow, staying clean becomes a delicate dance between what we need and what we have. It's a test of human resilience, proving that cleanliness doesn't have to be sacrificed to drought or the harsh winds of a dry land.

Maintaining personal hygiene with barely a drop to spare is a story of clever adjustments and innovative solutions, ensuring that disease doesn't find a foothold on our skin or spread through the community.

Prioritizing Hygiene Practices

When water is scarce, keeping clean becomes a strategic effort. We need to carefully choose our actions to protect ourselves from illness while respecting every precious drop. Handwashing takes center stage in this hierarchy.

A simple ritual, stripped down to its core, uses minimal water but gets a big boost from soap, becoming a powerful defense against spreading germs. Food handling also demands meticulous attention. We need to be extra careful to minimize contact with potential contaminants, making sure every bite nourishes instead of harms.

This focused approach, emphasizing the most crucial hygiene points, ensures we use water wisely without weakening the shield it offers against disease.

Water-Saving Hygiene Methods

In landscapes deprived of water, cleanliness finds new allies in water-efficient hygiene methods. These clever innovations stretch the usefulness of every drop, bending its purpose to the will of necessity. Sponge baths, an ancient practice reborn in times of scarcity, perfectly embody this idea.

With just a basin and a washcloth, we can cleanse our bodies, removing dirt and sweat with barely a splash of water. No-rinse cleansers, like baby wipes and hand sanitizers, are another testament to human ingenuity. These products offer cleaning power without the need for water. Their use is a reminder of the potential of chemistry to overcome the limitations of nature.

Your Essential Hygiene Kit

When every resource is carefully measured, putting together a hygiene kit becomes a strategic undertaking. This kit, a collection of tools to tackle the challenges of cleanliness during drought, might include:

- Microfiber Towels: These towels are highly absorbent and dry quickly, making them an efficient way to dry off after using minimal water.
- Biodegradable Soap: These soaps ensure that the little water you use doesn't harm the environment when disposed of. They dissolve completely, leaving no trace behind.
- Hand Sanitizer: A powerful ally in the fight against germs, hand sanitizer requires no water and offers significant protection.
- Dry Shampoo: This is a great option for maintaining hair care without the luxury of a full wash. It absorbs oil and refreshes the scalp, keeping you feeling clean.
- Wet Wipes: Pre-moistened wipes provide a way to cleanse your skin, surfaces, and utensils. They're a versatile tool in your hygiene arsenal.
- No-Rinse Cleansers: For those times when water feels like a distant memory, these cleansers offer a refreshing alternative to a bath. They remove odor and oil without the need for rinsing.

This kit is your armor against the challenges of staying clean, ensuring that even in the face of scarcity, cleanliness remains within reach.

Building a Culture of Cleanliness in a Community

Water scarcity can unravel a community as disease spreads more easily. Promoting hygiene within a community requires not just individual commitment but a collective awakening to the importance of cleanliness and health. Here are some strategies to foster this awakening:

- Hygiene Demonstrations: Visual and engaging tutorials that showcase water-efficient washing techniques can turn cleanliness into a shared ritual for the community.
- Shared Hygiene Stations: Equipped with essentials from the hygiene kit, these stations become hubs for cleanliness, points where the community comes together in the pursuit of health.
- Hygiene Ambassadors: These are individuals passionate about hygiene who share their knowledge within the community. They raise awareness and encourage people to follow good hygiene practices.
- Public Commitments to Cleanliness: A communal pledge, a vow to uphold hygiene standards, reinforces the idea that cleanliness is not just a personal responsibility but a shared commitment.

Through strategic prioritization, innovative hygiene methods, a well-equipped hygiene kit, and the fostering of a communal commitment to cleanliness, we can maintain good hygiene even in times of water scarcity.

12.4 WATER: THE HEARTBEAT OF SAFE AND SUSTAINABLE COOKING IN DRY PLACES

In regions with limited water, this essential resource becomes more than just a thirst quencher. It's a lifeline for safe food preparation and a key player in keeping our meals healthy. Water is involved in every step of cooking, from the very first wash of our ingredients to the final rinse of our dishes. But when water is scarce, we need to rethink our usual ways of doing things, finding methods that save water without compromising food safety.

Washing Wisely: Minimizing Water While Maximizing Cleanliness

Washing food is the foundation of clean cooking, but in dry areas, we have to get creative. Instead of dunking our vegetables in a big bowl of water, we can use a spray bottle to mist them with a fine layer. This gets rid of dirt and possible germs while using much less water. We can even use a vinegar solution on a clean cloth to wipe down fruits and vegetables. Vinegar is a natural disinfectant, so it helps keep food safe without needing a lot of rinsing.

Adapting Our Cooking Techniques: Embracing Water-Efficient Methods

Cooking itself traditionally uses a lot of water. Here too, we can adapt. Steaming is a great option because it recycles the water vapor. The steam condenses back into water that can be used for other cooking purposes. Another strategy is to choose dishes that require less water to begin with. Think stir-fries, grilled foods, or one-pot wonders! These methods might seem like adjustments forced by circumstance, but they actually show a deeper appreciation for water's importance in cooking. By using it wisely, we ensure there's enough for the most critical tasks.

Preventing Foodborne Illness: Careful Planning and Utensil Management

When water is limited, preventing the spread of germs from one food to another (cross-contamination) becomes even more important. Here, planning and careful handling of utensils become our best weapons.

Designate separate cutting boards for raw meat and vegetables. Even with limited water, make sure to clean them thoroughly between uses. This separation of tools and spaces helps keep food borne illnesses at bay, protecting the health of everyone eating the meal.

Water Conservation Beyond Cooking: Reusing for Cleaning

The water-saving approach doesn't stop after cooking. We can reuse the water we used for cooking (once it cools down, of course) to clean our pots and dishes. This water is already warm and soapy, making it a great way to cut through grease and grime. It's like getting double duty out of every drop!

Embracing Water-Conscious Cuisine: Planning Meals for Efficiency

Another way to save water is to plan meals that are naturally water-efficient. Think about dishes that require minimal washing of ingredients, or those that can be cooked using the same water for everything. One-pot meals are perfect examples!

By embracing these water-conscious cooking strategies, we not only conserve this precious resource, but we also open ourselves up to new culinary adventures.

A Blueprint for Sustainable Cooking: Lessons Beyond Dry Places

This journey into the many roles water plays in food preparation and safety, especially in dry areas, shows us a path of innovation and resourcefulness. It challenges us to rethink how we use water in the kitchen, not as something readily available, but as a valuable resource that deserves our respect and careful handling.

From using less water for washing and cooking to preventing cross-contamination and reusing water for cleaning, these strategies offer a blueprint for sustainable cooking practices.

These lessons aren't just for dry regions. They hold value for everyone who wants to conserve water and show respect for the natural world. By adopting these practices, we can all become more mindful cooks, ensuring safe and delicious meals while cherishing the precious resource that makes it all possible: water.

LEAVE A 1 CLICK REVIEW!

Customer Reviews

⭐⭐⭐⭐⭐ 2

5.0 out of 5 stars ▾

5 star	▓▓▓▓▓	100%
4 star		0%
3 star		0%
2 star		0%
1 star		0%

See all verified purchase reviews ›

Share your thoughts with other customers

Write a customer review ⬅

Since self-publishers really depend on readers' reviews, I would be incredibly grateful if you could take 60 seconds to write a brief review on Amazon, even if just a few sentences,

Please SCAN the QR CODE below to leave a review. Thank you!!

CONCLUSION

We've reached the end of our exploration together, and here's the most important reminder: water is essential for life. It's the foundation for everything we do, from staying healthy to simply surviving. This book has focused on the importance of water, especially in situations where you might not have easy access to it, like during emergencies or if you live "off the grid."

We've been on a journey of discovery together. We learned about dehydration and how much water we actually need every day. We explored different ways to collect water, from simple methods like catching rainwater to more advanced techniques like using filters and purifiers. We even looked at some fun do-it-yourself projects to help you become more self-sufficient when it comes to water.

The best part? These strategies can be used almost anywhere! Whether you live in a big city, a small town, or even in a place that experiences droughts or floods, the ideas in this book can be adapted to your situation.

This book is all about being prepared and learning as much as you can. By reading about water laws, the newest ways to purify water, and other important topics, you're taking a big step towards being ready for anything.

But just knowing things isn't enough. I want you to take action! Think about how prepared you are for water shortages right now. Choose the strategies that work best for you, and keep learning new skills to become even more water-savvy.

Remember, we're all in this together. Sharing your knowledge with others helps build strong communities that can face water challenges together. The more we work together, the better off we'll all be.

Writing this book was driven by a simple goal: to empower you. I want you to feel confident and capable of taking care of yourself and your loved ones when it comes to water. This book shouldn't just be a guide, it should be your trusted companion on your path to water security.

I applaud your dedication to learning about such a vital topic. Your effort shows how strong and committed you are to your own well-being and that of the people around you.

Let's look to the future with hope. With the knowledge and skills you've gained from this book, you'll feel prepared to manage your water resources, no matter what comes your way. Together, we can build a legacy of preparedness that will benefit generations to come.

Thank you for joining me on this journey. Here's to clean water and a brighter future for all!

REFERENCES

- Dehydration - Symptoms & causes - Mayo Clinic https://www.mayoclinic.org/diseases-conditions/dehydration/symptoms-causes/syc-20354086
- Water: How much should you drink every day? https://www.mayoclinic.org/healthy-lifestyle/nutrition-and-healthy-eating/in-depth/water/art-20044256
- Sustainable Water: How We Can Source Water More ... https://www.themomentum.com/ articles/sustainable-water-how-we-can-source-water-more-efficiently-and-lower-our-consumption
- Source Water Assessments | US EPA https://www.epa.gov/sourcewaterprotection/source-water-assessments#:~
- Harnessing the Sky: Innovative Rainwater Harvesting ... https://worldpermaculture association.com/rainwater-harvesting-8-methods/
- Rainwater Harvesting Laws You Need to Know About (2023) https://4perfectwater.com/
- EPA Home Water Testing Facts https://www.epa.gov/sites/default/files/2015-11/documents/2005_09_14_faq_fs_homewatertesting.pdf
- Myths About Drowning and Water Safety https://www.stopdrowningnow.org/myths-about-drowning-and-water-safety/
- How to Build a Rainwater Collection System: 13 Steps https://www.wikihow.com/Build-a-Rainwater-Collection-System
- Fog and Dew as Potable Water Resources https://www.ncbi.nlm.nih.gov/pmc/articles/ PMC7007155/
- How to Melt Ice and Snow to Find Drinking Water to Survive https://holtzmansurvival.com/blogs/news/how-to-melt-ice-and-snow-to-find-drinking-water-to-survive
- How Do Hydrologists Locate Groundwater? https://www.usgs.gov/special-topics/water-science-school/science/how-do-hydrologists-locate-groundwater
- Best Water Storage Containers for Emergencies [Tested] https://www.pewpewtactical. com/best-water-storage-containers/
- Creating and Storing an Emergency Water Supply https://www.cdc.gov/healthywater/ emergency/creating-storing-emergency-water-supply.html
- Rainwater Harvesting 101 | Your How-To Collect ... https://www.watercache.com/ education/rainwater-harvesting-101

- The problem of water caches on the PCT https://www.pcta.org/discover-the-trail/ backcountry-basics/leave-no-trace/problem-water-caches-pct/
- Boil Water Response-Information for the Public Health ... https://www.health.ny.gov/ environmental/water/drinking/boilwater/response_information_public_health_professional.htm
- DIY Charcoal Water Filters https://www.motherearthnews.com/sustainable-living/diy-charcoal-water-filters-zspz23112-zm0z23zatr0/
- CDC Solar Disinfection (SODIS) https://www.cdc.gov/safewater/publications_ pages/options-sodis.pdf
- UV Water Purification vs. Reverse Osmosis - Aqua Solutions https://www.aquasolutionspa.net/uv-water-purification-vs-reverse-osmosis#:~
- Detection of contaminants in water supply: A review on ... https://www.ncbi.nlm.nih.gov/pmc/articles/PMC7126548/
- Household Water Treatment https://www.cdc.gov/healthywater/global/household-water-treatment.html
- Heavy Metals in Water and Soil: Methods for Treatment https://mytapscore.com/blogs/tips-for-taps/heavy-metals-in-water-soil-methods-for-treatment
- How to Make a DIY Water Filtration System Using Sand or Gravel https://www.motherearthnews.com/diy/how-to-make-a-diy-water-filtration-system/
- DIY Water Distiller https://www.survivalresources.com/diy-water-distiller.html
- Rainwater Harvesting 101 | Your How-To Collect ... https://www.watercache.com/education/rainwater-harvesting-101
- Chapter: 8 Legal and Regulatory Issues https://nap.nationalacademies.org/read/21866/chapter/10
- Highly Efficient Water Harvesting with Optimized Solar ... https://onlinelibrary.wiley.com/doi/abs/10.1002/gch2.201800001
- Whose Water Is It Anyway? Comparing the Water Rights ... https://extension.okstate.edu/fact-sheets/whose-water-is-it-anyway.html#:~
- Environmental Benefits of Rainwater Harvesting & Rain ... https://www.bluebarrelsystems.com/blog/environmental-benefits-of-rainwater-harvesting/
- Rainwater Harvesting State Regulations and Technical ... https://www.pnnl.gov/main/publications/external/technical_reports/PNNL-24347Rev.1.pdf
- Sustainable Solutions: Discovering Rainwater Harvesting ... https://smartwateronline.com/news/rainwater-harvesting-techniques-used-worldwide#:~:

- Guidance for Responding to Drinking Water Contamination ... https://www.epa.gov/sites/default/files/2018-12/documents/responding_to_dw_contamination_incidents.pdf
- Water Treatment | Public Water Systems | Drinking Water https://www.cdc.gov/healthywater/drinking/public/water_treatment.html
- Food and Drinking Water Safety in a Radiation Emergency https://www.cdc.gov/nceh/radiation/emergencies/food_water_safety.html
- Drought Resilience and Water Conservation https://www.epa.gov/water-research/drought-resilience-and-water-conservation
- Chapter: 11 New and Emerging Drinking Water Treatment Technologies https://nap.nationalacademies.org/read/9595/chapter/13
- From seawater to drinking water, with the push of a button https://news.mit.edu/2022/portable-desalination-drinking-water-0428
- Five Ways Artificial Intelligence Is Shaping Future of Water ... https://www.wsp.com/en-us/insights/2023-artificial-intelligence-shaping-future-of-water
- Crowdsourcing and Citizen Science data for water ... https://www.space4water.org/news/crowdsourcing-and-citizen-science-data-water-resources-management
- Effects of Dehydration and Rehydration on Cognitive ... https://www.ncbi.nlm.nih.gov/pmc/articles/PMC6603652/
- Oral Rehydration Therapy - ORT: A Solution for Survival http://www.rehydrate.org/ors/solution-for-survival.htm
- 7 Most Common Waterborne Diseases (and How to Prevent Them) https://lifewater.org/blog/7-most-common-waterborne-diseases-and-how-to-prevent-them/
- Sodium for Survival: 8 Ways to Find This Essential Nutrient ... https://thegrownetwork.com/finding-sodium-wilderness-survival/
- Experts Name the Top 19 Solutions to the Global Freshwater Crisis https://www.circleofblue.org/2010/world/experts-name-the-top-19-solutions-to-the-global-freshwater-crisis/
- Off Grid Toilet A Comprehensive Guide https://theoffgridcabin.com/off-grid-toilet-a-comprehensive-guide/
- Survival Skills: 10 Ways to Purify Water https://www.outdoorlife.com/survival-skills-ways-to-purify-water/

- Tips for Handwashing When Running Water is Not Accessible https://www.usda.gov/media/blog/2020/05/21/tips-handwashing-when-running-water-not-accessible